Springer Theses

Recognizing Outstanding Ph.D. Research

Aims and Scope

The series "Springer Theses" brings together a selection of the very best Ph.D. theses from around the world and across the physical sciences. Nominated and endorsed by two recognized specialists, each published volume has been selected for its scientific excellence and the high impact of its contents for the pertinent field of research. For greater accessibility to non-specialists, the published versions include an extended introduction, as well as a foreword by the student's supervisor explaining the special relevance of the work for the field. As a whole, the series will provide a valuable resource both for newcomers to the research fields described, and for other scientists seeking detailed background information on special questions. Finally, it provides an accredited documentation of the valuable contributions made by today's younger generation of scientists.

Theses may be nominated for publication in this series by heads of department at internationally leading universities or institutes and should fulfill all of the following criteria

- They must be written in good English.
- The topic should fall within the confines of Chemistry, Physics, Earth Sciences, Engineering and related interdisciplinary fields such as Materials, Nanoscience, Chemical Engineering, Complex Systems and Biophysics.
- The work reported in the thesis must represent a significant scientific advance.
- If the thesis includes previously published material, permission to reproduce this must be gained from the respective copyright holder (a maximum 30% of the thesis should be a verbatim reproduction from the author's previous publications).
- They must have been examined and passed during the 12 months prior to nomination.
- Each thesis should include a foreword by the supervisor outlining the significance of its content.
- The theses should have a clearly defined structure including an introduction accessible to new PhD students and scientists not expert in the relevant field.

Indexed by zbMATH.

Michael Andrews

Search for Exotic Higgs Boson Decays to Merged Diphotons

A Novel CMS Analysis Using End-to-End Deep Learning

Doctoral Thesis accepted by
Carnegie Mellon University, Pittsburgh, USA

Springer

Author
Dr. Michael Andrews
Department of Physics
Carnegie Mellon University
Pittsburgh, PA, USA

Supervisor
Dr. Manfred Paulini
Professor of Physics
Carnegie Mellon University
Pittsburgh, PA, USA

ISSN 2190-5053 ISSN 2190-5061 (electronic)
Springer Theses
ISBN 978-3-031-25090-3 ISBN 978-3-031-25091-0 (eBook)
https://doi.org/10.1007/978-3-031-25091-0

This Springer imprint is published by the registered company Springer Nature Switzerland AG
The registered company address is: Gewerbestrasse 11, 6330 Cham, Switzerland

Supervisor's Foreword

The theory describing the smallest building blocks of matter and the forces acting between them is called the standard model of particle physics. It is an enormously successful theory describing the interactions of all known elementary particles, the quarks, leptons and gauge bosons acting as force mediators. Developed over the past 60 years, starting with the quark model in the 1960s, the discovery of the charm quark in 1974, the τ lepton seen in experiments from 1974 to 1977, the bottom quark in 1977, the W and Z bosons in 1983, the top quark in 1995, and culminating in the discovery of the Higgs boson in 2012, there is to date no experimental evidence contradicting the predictions of the standard model. Although it is successful in describing all phenomena at the subatomic scales, it is not a complete "theory of everything" that can explain all known observations. For example, no particle exists in the standard model that constitutes a possible candidate for the dark matter that makes up about one quarter of the energy–mass content of the universe. The quest for finding phenomena that are not described by the standard model is one reason why physicists at the CERN Large Hadron Collider (LHC) are searching for yet-unknown particles, which can pave the way to postulate theories beyond the standard model.

The Ph.D. research conducted by Dr. Michael Andrews under my supervision in the Department of Physics at Carnegie Mellon University using proton–proton collision data collected with the Compact Muon Solenoid (CMS) experiment at the LHC is not just another search for phenomena beyond the standard model. What sets the data analysis in Dr. Andrews' thesis apart from conventional CMS searches is the use of several innovative approaches and "firsts" for CMS. It is also a story about the beauty of being a professor that allows you to learn together with and from your students. Let's go back in time to better understand...

About five years I got interested in the application of modern machine learning (ML) techniques in particle physics. Somehow I had ignored ML for a long time given that we had been using neural networks in particle physics for over 30 years. Together with Dr. Andrews, I learned very quickly that recent ML advances, in particular in the field of computer vision, have led to breakthrough applications of convolutional neural networks to scientific challenges, if the data can be expressed as an image or series of images. In particular, we became interested in exploring whether ML can help to get beyond limitations of traditional analysis techniques. As a first project, Dr. Andrews' work demonstrated the application of image-based deep learning techniques to separate electron from photon showers in one of the CMS sub-detectors, the electromagnetic calorimeter, a task that is not achievable with conventional approaches. This brought us to establish the concept of an end-to-end event classification that directly leverages low-level detector data as input to classify event signatures such as using images from low-level detector data to go directly to classify an event signature without using data reconstruction.

Fueled by the initial success, Dr. Andrews became quite involved in ML and very quickly an expert in the usage of different deep learning networks and ML techniques. His thesis analysis follows the path of exploring what is the maximum information that can be extracted from detector data when modern ML approaches are unleashed. He studied the hypothetical decay of the Higgs boson into a pair of light particles $H \rightarrow \mathcal{A}\mathcal{A}$, each of which may in turn decay into a pair of photons $\mathcal{A} \rightarrow \gamma\gamma$. The branching fraction for $\mathcal{A} \rightarrow \gamma\gamma$ is maximized at light $m_{\mathcal{A}}$ masses, but in this regime, each of the $\mathcal{A} \rightarrow \gamma\gamma$ decays is highly merged, and the diphotons are reconstructed as a single photon shower in the CMS electromagnetic calorimeter consisting of lead-tungstate crystals. Using end-to-end ML techniques, Dr. Andrews was able to develop a mass regression algorithm that maintains sensitivity even in the limit, where the two photons from the $\mathcal{A} \rightarrow \gamma\gamma$ system deposit their energy in the same calorimeter crystal. On the way to setting the first CMS limit for the theoretically highly interesting mass regime $m_{\mathcal{A}} < 200$ MeV, Dr. Andrews solved several issues with sensitivity toward the $m_{\mathcal{A}} \rightarrow 0$ mass endpoint that I leave for the interested reader to discover in his thesis entitled "Search for exotic Higgs boson decays to merged photons employing a novel deep learning technique at CMS".

This well-written and nicely organized Ph.D. thesis contains very accessible introductions for the novice to particle physics but also allows the expert to find useful new information. For example, Chap. 2 is an engaging introduction to the LHC and the CMS detector that should be accessible for a reader less familiar with particle physics, while Chaps. 7 and 8 detail the mass regression method and the data analysis for the experts. There is something for everyone in this thesis.

Finally, let me conclude by expressing my appreciation for the Ph.D. thesis work of Dr. Michael Andrews. I feel honored and grateful about the opportunity that I had working with him and learning from him.

Pittsburgh, USA
December 2022

Dr. Manfred Paulini

Abstract

A search for exotic Higgs boson decays, of the form H→ aa, with a → $\gamma\gamma$, is performed. The hypothetical particle a is a light, scalar or pseudoscalar particle decaying to two highly merged photons reconstructed as a single photon-like object in the CMS detector. A novel, end-to-end deep learning-based technique is developed to directly measure the invariant mass of merged a → $\gamma\gamma$ candidates for the first time at CMS. Analysis criteria similar to those used in the standard model H→ $\gamma\gamma$ search are applied, to probe the possibility that existing measurements in this decay mode may conceal a contribution from a low-mass particle a. The search is performed using the full CMS Run II data set, corresponding to a total integrated luminosity of $136\,\text{fb}^{-1}$, at a proton–proton center-of-mass collision energy of $\sqrt{s} = 13\,\text{TeV}$. No significant excess of events over standard model expectations is found. Branching fractions for this process of $\mathcal{B}(\text{H} \rightarrow \text{aa} \rightarrow 4\gamma) \gtrsim 0.9\text{–}3.3 \times 10^{-3}$ are excluded at 95% confidence level, for particle masses between $0.1 \leq m_a \leq 1.2\,\text{GeV}$, assuming negligible lifetime.

Acknowledgments

I would like to express my profound gratitude to my adviser, Manfred Paulini, for his enduring and unwavering support over an admittedly longer-than-usual Ph.D. program. His faith and support have been instrumental in allowing me to explore riskier ideas that, while in retrospect have proven career-defining, could have easily been nipped at the bud during their earlier, more fragile days.

To the ECAL community for entrusting their cherished detector to my hands during Run II. Being ECAL run-coordinator has been the professional thrill of a lifetime. I can only hope that I have given back to them as much as they have given to me.

To Sergei Gleyzer, for his partnership in developing end-to-end ML. To my earlier advisers, Michael Wang, who gave me my big break into the world of high energy physics, and Neil Caranto, for prodding me to pursue physics when I was still a business graduate.

Finally, to my family for their love, moral (and financial) support, and to my friends at different stages of my Ph.D., for their company and day-to-day emotional support, wherever they may be today.

Contents

1 Introduction 1
References 3

2 The LHC and the CMS Detector 7
2.1 The LHC 7
2.2 Particle Interactions with Matter 13
2.3 The CMS Detector 14
2.3.1 Inner Tracking 16
2.3.2 Calorimeters 19
2.3.3 Magnet System 24
2.3.4 Outer Tracking 26
2.4 The CMS Trigger System 28
2.4.1 L1 Trigger 29
2.4.2 HLT Trigger 30
2.5 Physics Object Reconstruction 31
2.5.1 PF Elements 32
2.5.2 PF Objects 33
2.6 Detector Simulation 35
References 36

3 Theory and Phenomenology 39
3.1 The Standard Model 39
3.1.1 QED 42
3.1.2 QCD 43
3.1.3 Weak Interactions 45
3.1.4 Higgs Sector 46
3.2 Challenges with the SM 47
3.3 The Extended Higgs Sector 49
3.3.1 $\mathrm{H} \rightarrow \mathrm{aa} \rightarrow 4\gamma$ Phenomenology 52
References 56

4 Analysis Strategy ... 59
4.1 End-to-End Particle Reconstruction ... 59
4.2 $H \to aa \to 4\gamma$ Discrimination ... 60
4.3 Analysis Strategy ... 62
References ... 65

5 Data Sets ... 67
5.1 Recorded Data ... 67
5.2 Simulated Data ... 69
5.2.1 Signal Samples ... 69
5.2.2 Background Samples ... 73
References ... 75

6 Signal Selection ... 77
6.1 Trigger ... 78
6.2 Photon Preselection ... 79
6.3 $a \to \gamma\gamma$ Identification ... 80
6.4 Event Selection ... 81
6.5 Event Yields ... 82
References ... 83

7 $a \to \gamma\gamma$ Mass Regression ... 85
7.1 Image Construction ... 85
7.2 Training ... 86
7.3 Validation ... 90
7.4 Benchmarking ... 92
7.4.1 $a \to \gamma\gamma$ in Simulated $H \to aa \to 4\gamma$ ... 92
7.4.2 $\pi^0 \to \gamma\gamma$ in Data ... 95
7.5 Robustness of the Algorithm ... 98
7.5.1 $\pi^0 \to \gamma\gamma$ in Data ... 98
7.5.2 e^+e^- Electrons in Data ... 100
7.5.3 $Z\ e^+e^-$ Electrons in Data Versus Simulation ... 101
7.5.4 $a \to \gamma\gamma$ in Simulated $H \to aa \to 4\gamma$ ... 103
References ... 106

8 Analysis ... 107
8.1 Signal Model ... 107
8.1.1 Photon Identification Scale Factors ... 109
8.1.2 HLT Trigger Scale Factors ... 114
8.1.3 Pileup Re-weighting ... 114
8.1.4 Sample Interpolation ... 115
8.2 Background Model ... 119
8.2.1 QCD Jet Substructure ... 121
8.2.2 Background Estimation ... 122
8.2.3 Background Validation ... 130
8.3 Systematics ... 136
8.3.1 p_T Re-weighting ... 137

8.3.2 Background Parametrization ... 139
8.3.3 $H \rightarrow \gamma\gamma$ Template Fraction ... 140
8.3.4 Luminosity Uncertainty ... 142
8.3.5 Photon ID Scale Factor Uncertainty ... 142
8.3.6 HLT Trigger Scale Factor Uncertainty ... 143
8.3.7 m_Γ Regressor Scale and Smearing ... 143
8.3.8 Summary ... 144
References ... 146

9 Results ... 147
9.1 Expected Sensitivity ... 147
9.1.1 Maximum Likelihood Estimation (MLE) ... 148
9.1.2 Impacts ... 151
9.1.3 Control Region Signal Extraction Tests ... 153
9.1.4 N-1 Expected Upper Limits ... 155
9.1.5 Expected Upper Limits ... 157
9.2 Observed Results ... 159
9.2.1 Impacts ... 159
9.2.2 Goodness-of-fit ... 160
9.2.3 Observed 2D-m_Γ Distribution ... 161
9.2.4 Observed Upper Limits ... 162
9.2.5 Discussion ... 169
References ... 171

10 Conclusions ... 173
References ... 177

Appendix: Supplementary Studies ... 179

Chapter 1
Introduction

In 2012, a new boson was observed by the CMS and ATLAS experiments [1, 2] operating at the CERN Large Hadron Collider (LHC) with properties consistent with the standard model (SM) Higgs boson decaying to $\mathrm{H} \rightarrow \mathrm{ZZ}^* \rightarrow 4\ell$ and $\mathrm{H} \rightarrow \gamma\gamma$. Since then, additional decay modes have been observed, building confidence that the new boson is, in fact, the SM Higgs boson [3, 4], capping off a major puzzle in the origin of electroweak symmetry breaking and particle mass. Indeed, the results of the broader CMS search program of recent years suggest that the physics probed by the LHC is just as predicted by the SM. Yet, astronomical observations and theoretical inconsistencies [5] make it clear that the SM cannot be the final theory of particle physics. With the LHC now collecting unprecedented amounts of data, this has prompted a number of searches for physics beyond the standard model (BSM) that venture farther out into unexplored corners of phase space, where they may have been overlooked by more conventional search strategies.

Even in the light of current LHC constraints, the Higgs sector remains an important search space for BSM physics, due to its accessibility to SM-neutral hidden sectors. In such scenarios, because of the small decay width of the SM Higgs boson, even minute couplings to BSM physics can lead to sizeable branching fractions for exotic new states that may be accessible at the LHC. With current constraints on $\mathrm{H} \rightarrow \mathrm{BSM} \approx 20-60\%$ [6], depending on assumptions, this leaves much room for exploration in the exotic Higgs sector.

At the same time, recent advances in analysis tools, particularly those based on advanced machine learning (ML) or so-called deep learning, have empowered the pursuit of experimentally challenging topologies, which were theoretically well-motivated but simply not feasible to pursue previously. A prime example, which is the focus of this thesis, is the exotic decay of the Higgs boson to a pair of light scalars, each subsequently decaying to two photons, $\mathrm{H} \rightarrow \mathrm{aa}$ with $\mathrm{a} \rightarrow \gamma\gamma$ [7], or $\mathrm{H} \rightarrow \mathrm{aa} \rightarrow 4\gamma$ for short. Not all applications of ML, however, lead to breakthroughs. Historically, the LHC experiments used highly processed inputs representing physically meaningful quantities like particle 4-momenta to train ML algorithms. However, for many new physics searches, sensitivity is limited not by the ability to extract useful information from particle 4-momenta but by inefficiencies in the reconstruction of the 4-momenta

M. Andrews, *Search for Exotic Higgs Boson Decays to Merged Diphotons*, Springer Theses, https://doi.org/10.1007/978-3-031-25091-0_1

quantities themselves. In this thesis, we describe the first LHC physics analysis that trains advanced ML algorithms directly on low-level, "raw" detector data to address this bottleneck. As we will show, a direct probe of the $\mathrm{H} \rightarrow \mathrm{aa} \rightarrow 4\gamma$ signal at CMS is impossible without end-to-end deep learning. Thus, for suitable applications, end-to-end deep learning delivers breakthrough sensitivity and extends our physics reach to entirely new domains.

Decays like those of $\mathrm{H} \rightarrow \mathrm{aa} \rightarrow 4\gamma$ are candidates in various BSM models involving Minimal Composite Higgs Models (MCHM), two-Higgs-double-like models (2HDM), Next-to-Minimal Supersymmetric Standard Model (NMSSM), and any SM extension involving an additional hidden sector coupling to a new singlet [8, 9]. Moreover, such decays are of particular interest in searches for axion-like particle (ALP) production [10–13] because of their potential impact on our understanding of the early universe and stellar formation. ALPs are also an important potential candidate for dark matter [14–17]. In astrophysical and cosmological searches, the particle a is additionally identified as a spin-0 or CP-odd particle, known as a pseudoscalar. The experimental search we present, however, is insensitive to the CP quantum numbers of a, since its polarization is not measured.

While different model assumptions allow for varying $\mathrm{a} \rightarrow \gamma\gamma$ branching fractions, the $\mathrm{a} \rightarrow \gamma\gamma$ decay mode is generally enhanced when m_{a} is less than the pair production threshold for decays to the heavier SM states [8]. For masses below the charmonium production threshold ($m_{\mathrm{a}} \lesssim 3\,\mathrm{GeV}$), the particle a will be increasingly preferred to be long-lived [8]. If the a decays outside of the detector volume, it will not be reconstructed at all. Moreover, even if the a decays promptly, if it arise from $\mathrm{H} \rightarrow \mathrm{aa}$, the $\mathrm{a} \rightarrow \gamma\gamma$ photons will be highly collimated. Each $\mathrm{a} \rightarrow \gamma\gamma$ will thus be misreconstructed as a single photon-like object $\Gamma(\mathrm{a} \rightarrow \gamma\gamma)$, or Γ for short, by existing particle reconstruction algorithms. In this scenario, the $\mathrm{H} \rightarrow \mathrm{aa} \rightarrow 4\gamma$ decay will present an invariant mass resonance approximately degenerate with that of the SM $\mathrm{H} \rightarrow \gamma\gamma$ decay [18]. Therefore, if realized in nature, the low-m_{a} $\mathrm{H} \rightarrow \mathrm{aa} \rightarrow 4\gamma$ signal will be buried in existing events resembling SM $\mathrm{H} \rightarrow \gamma\gamma$ decays [6, 19].

Motivated by these challenges and opportunities, in this thesis, we present the first $\mathrm{H} \rightarrow \mathrm{aa} \rightarrow 4\gamma$ search that directly measures the invariant mass spectrum of merged photon candidates Γ reconstructed in events resembling a SM $\mathrm{H} \rightarrow \gamma\gamma$ final state. That is, the search is performed in the experimentally challenging regime where the $\mathrm{a} \rightarrow \gamma\gamma$ decays are merged, but where the branching fraction for this decay mode is most theoretically attractive. The analysis is made possible by the development of a novel particle reconstruction technique, which we likewise describe in this thesis. The technique utilizes an end-to-end deep learning strategy to reconstruct the invariant mass of merged photon candidates, m_{Γ}, directly from the energy deposits in the CMS electromagnetic calorimeter. The full CMS Run II data set is used, corresponding to an integrated luminosity of 136 fb^{-1}. We probe $\mathrm{H} \rightarrow \mathrm{aa} \rightarrow 4\gamma$ decays with particle a masses in the range range $m_{\mathrm{a}} = 0.1$–1.2 GeV. In this first analysis, we assume that the as decay promptly and analyze only $\mathrm{a} \rightarrow \gamma\gamma$ candidates reconstructed in the barrel section of the detector, for simplicity.

While a number of ATLAS measurements have performed similar searches [18, 20], this analysis represents the first attempt at the LHC to directly probe the $\mathrm{a} \rightarrow \gamma\gamma$

invariant mass spectrum. A number of other CMS analyses have been published [21–25], or are underway, to either directly or indirectly search for particle a decays to other states $a \to xx$, as well its possible production from yet another new state, $X \to aa$. Generic decays of the form $a \to \gamma\gamma$ have also been studied outside of $H \to aa$ decays in collider experiments [26, 27], as well as in astrophysics and cosmology [5, 28, 29], although at much lighter masses $m_a \sim$ eV.

This thesis is based on two CMS results. The first of these is a technique paper [30] describing the development and validation of the end-to-end deep learning technique in the context of $a \to \gamma\gamma$ decays. The second of these is a physics analysis [31] focusing on the application of this technique to perform the first direct search for $H \to aa \to 4\gamma$ in its most experimentally challenging but theoretically attractive regime. It will take many years to bear out the ultimate value of the end-to-end deep learning to physics searches but this work represents an important first step in demonstrating its potential. Already, entirely new tools and searches at CMS are underway that push the limits of what can be probed with the CMS detector using end-to-end deep learning at their core. The lasting significance of this work, therefore, will arguably be its demonstration of the feasibility and breakthrough potential of the end-to-end deep learning technique for physics searches.

This thesis is arranged as follows. Following this chapter which describes the motivation for the $H \to aa \to 4\gamma$ search, a description of the CERN LHC experimental apparatus and the CMS detector collecting the data is provided in Chap. 2. The theoretical basis of the SM, the extended Higgs sector, and the phenomenology of the $H \to aa \to 4\gamma$ decay are then presented in Chap. 3. In Chap. 4, we outline the analysis strategy for discriminating $H \to aa \to 4\gamma$ signal events. The CMS data sets used for the analysis, and the criteria employed to select $H \to aa \to 4\gamma$ candidate events, are detailed in Chaps. 5 and 6, respectively. Chap. 7 is dedicated to describing the training and validation of the novel end-to-end ML-based m_Γ regression algorithm. The main physics analysis, detailing the signal and background models used to perform the $H \to aa \to 4\gamma$ signal search, is given in Chap. 8. The results of the analysis are presented in Chap. 9, and our conclusions are summarized in Chap. 10.

References

1. Chatrchyan S, Khachatryan V, Sirunyan AM, Tumasyan A, Adam W, Aguilo E, Bergauer T, Dragicevic M, Ero J, Fabjan C et al (2012) Observation of a new boson at a mass of 125 GeV with the CMS experiment at the LHC. Phys Lett B 716:30 Sep
2. Aad G, Abajyan T, Abbott B, Abdallah J, Abdel Khalek S, Abdelalim A, Abdinov O, Aben R, Abi B, Abolins M et al (2012) Observation of a new particle in the search for the standard model Higgs boson with the atlas detector at the LHC. Phys Lett B 716:1
3. Sirunyan AM, Tumasyan A, Adam W, Ambrogi F, Asilar E, Bergauer T, Brandstetter J, Dragicevic M, Ero J, Escalante Del Valle A et al (2018) Observation of Higgs boson decay to bottom quarks. Phys Rev Lett 121

4. Aaboud M, Aad G, Abbott B, Abdinov O, Abeloos B, Abhayasinghe D, Abidi S, AbouZeid O, Abraham N, Abramowicz H et al (2018) Observation of Higgs boson production in association with a top quark pair at the LHC with the atlas detector. Phys Lett B 784:173 Sep
5. Zyla P et al (2020) Review of particle physics. PTEP 2020(8):083C01
6. Sirunyan AM, Tumasyan A, Adam W, Ambrogi F, Asilar E, Bergauer T, Brandstetter J, Dragicevic M, Ero J, Del Valle AE et al (2019) Combined measurements of Higgs boson couplings in proton-proton collisions at $\sqrt{s} = 13$ TeV. Eur Phys J C 79:421
7. Dobrescu BA, Landsberg G, Matchev KT (2001) Higgs boson decays to CP-odd scalars at the Fermilab Tevatron and beyond. Phys Rev D 63:075003 Feb
8. Curtin D, Essig R, Gori S, Jaiswal P, Katz A, Liu T, Liu Z, McKeen D, Shelton J, Strassler M et al (2014) Exotic decays of the 125 GeV Higgs boson. Phys Rev D 90
9. Curtin D, Essig R, Gori S, Shelton J (2015) Illuminating dark photons with high-energy colliders. J High Energy Phys 157
10. Peccei RD, Quinn HR (1977) CP conservation in the presence of pseudo particles. Phys Rev Lett 38:1440–1443 Jun
11. Bauer M, Neubert M, Thamm A (2017) Collider probes of axion-like particles. JHEP 12:044
12. Peccei RD (2008) The strong CP problem and axions. Springer, Berlin Heidelberg, p 3
13. Essig R et al (2013) Dark sectors and new, light, weakly-coupled particles
14. Raffelt GG (2008) Astrophysical axion bounds. Springer, Berlin Heidelberg, p 51
15. Sikivie P (2008) Axion cosmology. Springer, Berlin Heidelberg, p 19
16. Marsh DJ (2016) Axion cosmology. Phys Rep 643:1
17. Chadha-Day F, Ellis J, Marsh DJE (2021) Axion dark matter: what is it and why now?
18. ATLAS Collaboration (2012) Search for a Higgs boson decaying to four photons through light CP-odd scalar coupling using 4.9 fb^{-1} of 7 TeV pp collision data taken with ATLAS detector at the LHC
19. Khachatryan V et al (2014) Observation of the diphoton decay of the Higgs boson and measurement of its properties. Eur Phys J C 74(10):3076
20. Aaboud M et al (2019) Search for pairs of highly collimated photon-jets in pp collisions at $\sqrt{s} = 13$TeV with the ATLAS detector. Phys Rev D 99:012008 Jan
21. Sirunyan AM, Tumasyan A, Adam W, Ambrogi F, Asilar E, Bergauer T, Brandstetter J, Dragicevic M, Ero J et al (2018) Search for an exotic decay of the Higgs boson to a pair of light pseudoscalars in the final state of two muons and two τ leptons in proton-proton collisions at $\sqrt{s} = 13$ TeV. J High Energy Phys 2018
22. Sirunyan AM, Tumasyan A, Adam W, Ambrogi F, Asilar E, Bergauer T, Brandstetter J, Dragicevic M, Ero J, Escalante Del Valle A et al (2019) A search for pair production of new light bosons decaying into muons in proton-proton collisions at 13 TeV. Phys Lett B 796:131 Sep
23. Sirunyan AM, Tumasyan A, Adam W, Ambrogi F, Asilar E, Bergauer T, Brandstetter J, Dragicevic M, Ero J, Escalante Del Valle A et al (2020) Search for light pseudoscalar boson pairs produced from decays of the 125 GeV Higgs boson in final states with two muons and two nearby tracks in pp collisions at $\sqrt{s} = 13$ TeV. Phys Lett B 800:135087 Jan
24. Sirunyan AM, Tumasyan A, Adam W, Ambrogi F, Asilar E, Bergauer T, Brandstetter J, Dragicevic M, Ero J, Escalante Del Valle A et al (2019) Search for an exotic decay of the Higgs boson to a pair of light pseudoscalars in the final state with two muons and two b quarks in pp collisions at 13 TeV. Phys Lett B 795:398 Aug
25. Sirunyan AM, Tumasyan A, Adam W, Ambrogi F, Bergauer T, Dragicevic M, Ero J, Escalante Del Valle A, Fruhwirth R et al (2020) Search for a light pseudoscalar Higgs boson in the boosted $\mu\mu\tau\tau$ final state in proton-proton collisions at $\sqrt{s} = 13$ TeV. J High Energy Phys 2020
26. Knapen S, Lin T, Lou HK, Melia T (2017) Searching for axionlike particles with ultraperipheral heavy-ion collisions. Phys Rev Lett 118:171801
27. Adhikari S et al (2022) Search for photoproduction of axionlike particles at GlueX. Phys Rev D 105:052007
28. Graham PW, Irastorza IG, Lamoreaux SK, Lindner A, van Bibber KA (2015) Experimental searches for the axion and axion-like particles. Annu Rev Nucl Part Sci 65:485

29. Irastorza IG, Redondo J (2018) New experimental approaches in the search for axion-like particles. Prog Part Nucl Phys 102:89
30. Reconstruction of decays to merged photons using end-to-end deep learning with domain continuation in the CMS detector (2022)
31. Search for exotic Higgs boson decays $\mathrm{H} \to \mathcal{A}\mathcal{A} \to 4\gamma$ with events containing two merged diphotons in proton-proton collisions at $\sqrt{s} = 13$ TeV (2022)

Chapter 2
The LHC and the CMS Detector

In this chapter, we describe the experimental apparatus involved in the production, collection, and reconstruction of the particle physics data used in this analysis. The basic unit of statistically independent physics data is the collisions event, or event for short. In Sect. 2.1, we begin with a description of the Large Hadron Collider (LHC), which is the primary apparatus responsible for the production of high energy collision events. This is followed in Sect. 2.3 by a description of the Compact Muon Solenoid (CMS) detector, which is responsible for the collection of data generated by the LHC, and the main data source for this analysis. A short primer on the interaction of particles with matter is presented in Sect. 2.2, prior to the description of the CMS detector, in order for the design of the CMS detector to be better appreciated. Following these, the steps involved in the filtering and reconstruction of the detector data are described. Due to the untenable volume of data generated by the LHC, a dedicated event filtering or *triggering* system is implemented in the CMS detector, to select only events of interest, described in Sect. 2.4. For events passing the trigger, the data collected from the CMS subdetectors are subsequently reconstructed into physics objects used for analysis, as described in Sect. 2.5. Note that the reconstruction here pertains to those of standard CMS physics objects, not those reconstructed by the end-to-end technique, which is instead described in Chap. 7. Finally, as particularly relevant for the end-to-end reconstruction technique, we conclude this chapter in Sect. 2.6 with an overview of the detector simulation process and its basic validation.

2.1 The LHC

The CERN LHC is presently the largest and most energetic man-made particle collider ever built. It straddles the border of France and Switzerland, between the foothills of the Jura mountain range and Lac Léman, some 100 km underground.

The LHC, while designed to be a general purpose collider, was conceived primarily to study the nature of electroweak symmetry breaking, for which the Higgs mechanism was thought to be responsible. Today, it is chiefly known for its discov-

M. Andrews, *Search for Exotic Higgs Boson Decays to Merged Diphotons*,
Springer Theses, https://doi.org/10.1007/978-3-031-25091-0_2

ery of the Higgs boson, jointly discovered by the CMS and ATLAS experiments in their Run I (2011–2012) data sets, for which the 2013 Nobel prize in physics was awarded to the duo of Francois Englert and Peter Higgs. The LHC remains the only operational collider able to probe the electroweak energy regime and thus continues to host a broad research program investigating both high-precision, high-energy SM physics, as well as searches for physics beyond the standard model.

In this section, we describe the design choices that motivated the LHC's construction, detail its basic operation, and highlight key features that drive its physics performance.

Collider Design. The LHC, at its most basic level, is a synchrotron accelerator that accelerates beams of charged particles in a circular orbit. In the case of the LHC, there are two counter-rotating beams of protons, which, at pre-determined points in the orbit, are steered into collision, from which particle collisions are generated.

As opposed to linear accelerators, colliders based on circular accelerators have the distinct advantage of having much higher collision rates. At an energy of 6.5 TeV per beam, each proton in the beam orbits the 27 km circumference of the LHC ring at a rate of more than 11 kHz, orders-of-magnitude higher than would be achievable with linear accelerators that would need to be time-consumingly refilled after every collision. As a result, the LHC has one of the highest nominal collision rates of any collider, 40 MHz, placing it in a unique position to probe the rarest of physics decays.

As opposed to striking fixed targets, by introducing two counter-rotating beams, the LHC is additionally able to maximize collision energy. For a given beam energy, the collision energy, parametrized by the Mandelstam variable $\sqrt{s}$, is maximized when the incident particles collide in their center-of-mass frame. By utilizing two counter-rotating beams of similar mass and energy, the physics potential of the LHC beams is therefore maximized in the lab frame. As a result, the LHC is also the most energetic collider ever built, with a $\sqrt{s} = 13$ TeV, giving it the ability to probe the highest energy physical phenomena, or equivalently, the smallest length scales, in a laboratory setting.

A disadvantage of circular colliders, however, is that they require magnets with large bending power in order to deflect particles into a circular orbit. For an orbit radius R, a particle of charge q and momentum $|\mathbf{p}|$ requires a magnetic field of strength

$$|\mathbf{B}| = \frac{|\mathbf{p}|}{qcR}, \tag{2.1}$$

where c is the speed of light. Moreover, accelerating charged particles (as in a circular orbit) dissipate energy in the form synchrotron radiation. For a particle of mass m, this occurs at a rate proportional to

$$P \propto \frac{q^2}{m^4 R^2} \tag{2.2}$$

As the above equations suggest, these effects can be mitigated by constructing large radius accelerator rings and using heavy particles. It should come as no surprises then that the LHC is also the largest collider to have ever been built, with a radius of $R = 4.3$ km. In a previous incarnation, known then as the Large Electron Positron (LEP) collider, electrons were originally collided. However, with an upgraded ring designed for bending protons, the new ring came to be known as the LHC.

Choosing to collide protons, a kind of hadron, however, has its drawbacks, and for which the LEP was first constructed. Unlike electrons, protons are not point particles, but rather bound states of quarks and gluons, or partons. Because of this, the full $\sqrt{s} = 13$ TeV of the proton collision is not actually transferred to the primary interacting partons. Only a fraction of this energy, dictated by the proton's parton distribution function, is at any time available to the partons interacting in the primary collision event, known as the hard scattering event. Indeed, because there are other partons that share in the proton's energy, they too may collide in softer scattering events, known as the underlying event. Moreover, partons, and in the case of softer collisions, other hadrons, interact predominantly through QCD. Thus, while the LHC is a prodigious collision factory, majority of these events do not directly illuminate electroweak or BSM physics.

On balance, however, the LHC succeeds as a general purpose collider. The LHC has managed to "re-discover" all the major SM particles discovered by its predecessors, in addition to singularly discovering a new boson consistent with the SM Higgs boson. As of this writing, plans are being drawn up to build a successor to the LHC, currently dubbed the Future Circular Collider (FCC), that pushes the same circular collider strategy of the LHC to new heights.

The LHC Complex. As with all synchrotrons, the LHC bends charged particle beams through the use of magnets. For the demands of the LHC, superconducting magnets are utilized. Each segment of the LHC ring is composed of cryogenically-cooled superconducting dipole magnets, housing the two counter-rotating proton beam pipes, to regularly deflect the particle beams into the desired, approximately circular orbit. Interspersed between the superconducting dipoles are superconducting quadrupoles which behave like lenses for beam focusing, and other, high-order moment magnets for various beam correction operations. The magnets are constructed from niobium-titanium (NbTi) which, during operation, produce a field strength of 7.7 T. To keep the magnets in a superconducting state, they are cooled to 1.9 K using liquid helium.[1]

The LHC is but the last stage of a larger LHC accelerator complex, as illustrated in Fig. 2.1. Before the proton beams are accelerated by the LHC to an energy of $E = 6.5$ TeV, they are sequentially accelerated through a series of linear and circular accelerators, each optimized to bring the beams to progressively higher energies. First, the protons start out as molecular hydrogen (H_2) in a compressed gas tank. Hydrogen ions (H^-) are then extracted and injected into a linear accelerator (Linac4). The Linac4 uses radio frequency (RF) cavities to accelerate the hydrogen ions to 160

[1] A major accident occurred in 2008 in which one of the superconducting LHC magnets "quenched" or lost its superconducting state, resulting in rapid heating and expansion of the liquid helium coolant, enough to cause a small explosion.

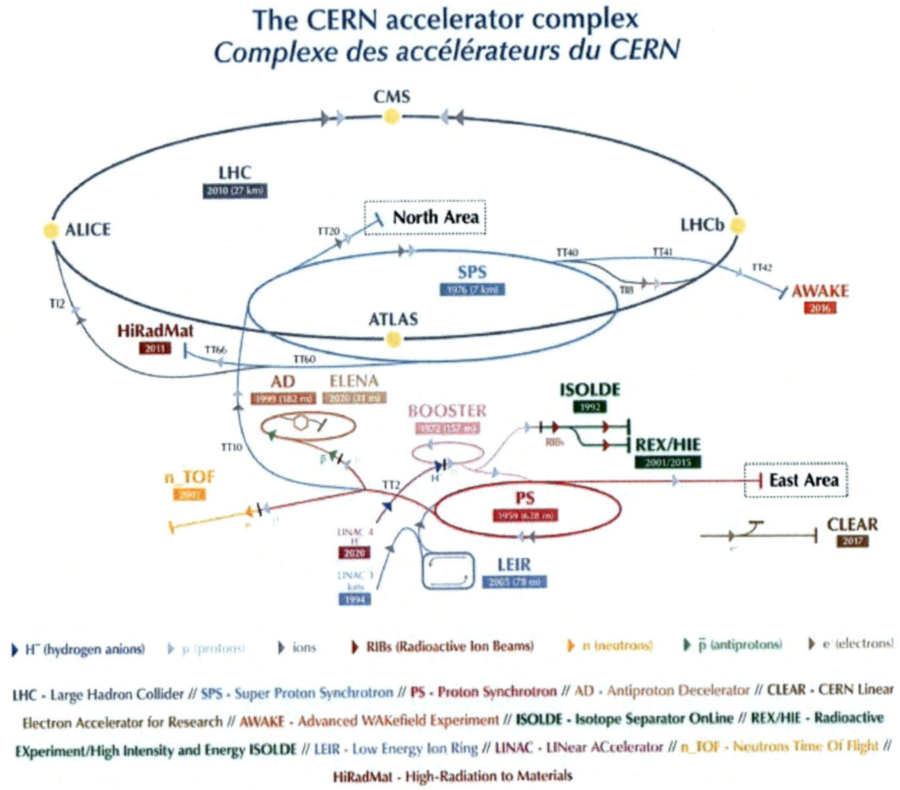

Fig. 2.1 The full LHC accelerator complex. *Credit* CERN

MeV. The hydrogen ions are then injected into the first circular accelerator, the Proton Synchrotron Booster (PSB). The PSB strips off the electrons from the hydrogen ion beams, leaving only protons, which are accelerated to 2 GeV. The resulting proton beam is then injected into the Proton Synchrotron (PS), where they are further accelerated to 26 GeV, before being injected into the Super Proton Synchrotron (SPS). The SPS brings the proton beams to an energy of 450 GeV before, finally, injection into the main LHC ring. To insulate the LHC ring from surface vibrations that could disturb the proton beam alignment, the LHC ring is situated 100 km underground.

In the LHC ring, the incoming proton beam is accumulated into distinct bunches, before being accelerated to the full 6.5 TeV energy. In total, each of the two LHC beams consists of more than 2000 bunches, organized into bunch "trains", with each bunch containing about 10^{11} protons. The entire injection sequence, from hydrogen gas to filled proton bunches in the LHC, takes about two hours.

Once the proton beams in the LHC are ramped to their full operating energy, the counter-rotating beams a gradually steered into collision at four "interaction points", spaced roughly symmetrically around the LHC ring. At a nominal bunch spacing of 25 ns, this results in bunches crossing at a rate of 40 MHz at each of the four points.

The timing of the bunches is closely controlled and communicated to the individual LHC experiments to allow them to precisely synchronize their data collection.

It is around these four interaction points that the four main LHC detectors are built: ATLAS, CMS, LHCb, and ALICE. ATLAS and CMS are both general purpose detectors with overlapping physics objectives. LHCb primarily studies CP violation using b-quark decays, while ALICE studies the quark-gluon plasma. There are four additional, ancillary experiments that operate off the four interaction points: TOTEM, LHCf, MoEDAL, and FASER.

For the LHC ring, the main constraint preventing higher beam energies (and thus collision energies) is not radiative losses ($\sim$ 1 keV per orbit) but the bending power of the dipole magnets. Following the quench incident of 2008 (see footnote 1) the LHC magnets had to be operated at significantly lower field strengths, resulting in the LHC collision energy being halved.

Physics Performance. While the LHC proton bunches cross at 40 MHz, the actual rate at which events are produced for a given physics process is a function of the beam luminosity L. For a physics process with an LHC production cross section of σ_{XS}, the event production rate is

$$N_{\mathrm{events}} = L\sigma_{\mathrm{XS}}, \tag{2.3}$$

where L is proportional to the number and Lorentz boost of the protons in the beam, and inversely proportional the geometric cross section of the beam (i.e., the more focused the beam, the higher the luminosity).

As listed in Fig. 2.2, the overwhelming majority of events produced by the LHC are soft QCD interactions, with the contribution from electroweak physics exceedingly rare. The biggest contributor to the soft QCD interactions is the non-diffractive inelastic scattering, or so-called minimum bias events, with a production cross section of about 65 mb $\sim 10^{10}$pb. During the recently completed Run II phase (years 2016–2018) of the LHC, where the beam luminosity typically peaked at about $L \sim 2 \times 10^{34}\,\mathrm{cm}^{-2}\,\mathrm{s}^{-1} \sim 2 \times 10^{-2}\mathrm{pb}^{-1}\,\mathrm{s}^{-1}$ at the start of the proton beam fill, this translates to about $\sim 10^9$ minimum bias events per second.[2] By contrast, for SM Higgs production, with a cross section of about $\sigma_H \sim 50$ pb, only $\sim$ 1 Higgs boson is produced every second.

During operation, the LHC collision rate is such that a single proton beam fill can last for 12 h or longer before being dumped, over which time the luminosity typically drops by half or more.

The proton bunching scheme has important consequences on physics performance. In addition to the underlying event described earlier due to the use of protons, having 10^{11} protons in each colliding bunch results in the possibility of multiple, simultaneous interactions per bunch crossing. This phenomenon, known as pileup, can result in additional soft-scattering events being overlaid onto the primary hard-

[2] $1\,\mathrm{cm}^2 = 1 \times 10^{24}$ pb.

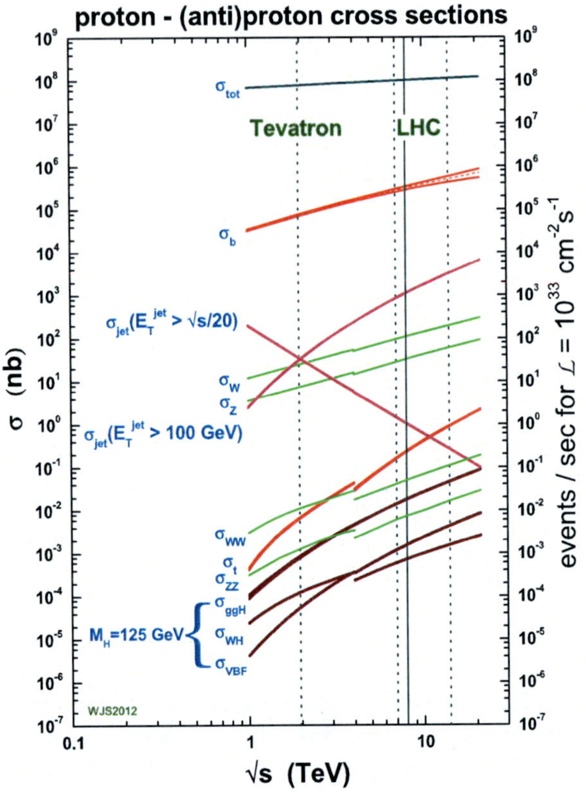

Fig. 2.2 Total physics production cross section at the LHC relative to component from QCD jet physics, Higgs boson production, and other electroweak processes. *Credit* W.J. Stirling, Imperial College London

scattering event,[3] complicating attempts to reconstruct the primary collision event. While this is mitigated by the identification of the primary collision vertex (see Sect. 2.5), it can nonetheless lead to spurious, soft particle tracks and energy deposits in the detector.

At CMS, one typically acknowledges two sources of pileup: in-time pileup, where the additional interaction arise from other protons in the same bunch crossing as the primary collision, and out-of-time pileup, where they arise from protons in a prior or later bunch crossing. During Run II, the mean pileup level measured at the CMS interaction point, as shown in Fig. 2.3, was between 40 and 60 additional interactions, in proportion to the instantaneous beam luminosity. Notably, during 2017, an issue with the LHC magnet cooling resulted in the proton beams being switched to an alternate bunching scheme that resulted in higher out-of-time pileup contributions. These are visible in the middle plot in Fig. 2.3 as a secondary peak at higher pileup values for 2017.

[3] In practice, multiple hard-scattering events could also occur, but the rate for this to occur would be doubly rare.

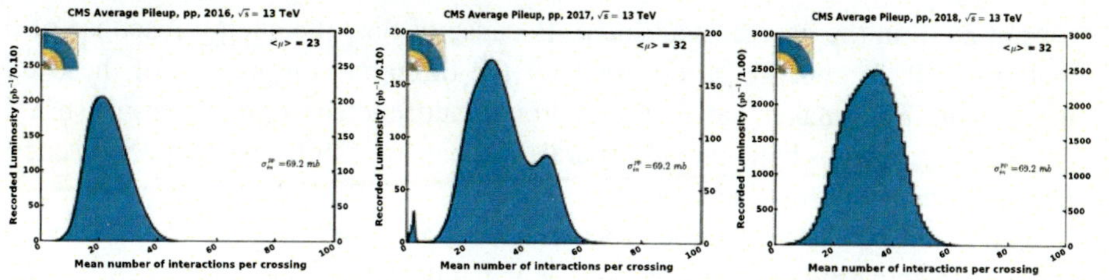

Fig. 2.3 Mean number of interactions per bunch crossing for the proton-proton beams in the LHC for 2016 (left), 2017 (middle), and 2018 (right) data-taking periods. *Credit* CMS

2.2 Particle Interactions with Matter

Before proceeding to describe the CMS detector, it is helpful to review the physics behind particle interactions in matter, as they represent a major driving factor in the design of the individual CMS subdetectors.

Electromagnetic Showers. At high energies, electrons and photons incident upon some material typically dissipate energy through electromagnetic interactions with the atomic nuclei of the material. For photons with energies above the e^+e^- pair production threshold, incident photons predominantly interact electromagnetically with the atomic nuclei to produce an outgoing e^+e^-. For high energy electrons (or positrons), these again interact electromagnetically with the nuclei to radiate a bremsstrahlung photon. Thus, regardless of whether the original incident particle was an electron or a photon, a cascade of electromagnetic particles is generated in the material. This "electromagnetic shower" continues until the pair production threshold is reached, at which point, lower energy dissipation mechanisms such as ionization proceed. In the process, the energy of the original incident electromagnetic (EM) particle is absorbed into the material, an effect that can be exploited to build electromagnetic calorimeters for measuring the energy of EM particles.

The radiation length X_0 of a material is the characteristic distance over which a high energy electron interacting with the material dissipates all but e^{-1} of its original energy. It is equal to $7/9$ of the mean free path for a high energy photon interacting with the same material to decay into an e^+e^- pair. The Molière radius R_M of a material, a related parameter, gives the characteristic energy dissipation length perpendicular to the incident particle's trajectory: approximately 90% of an incident high-energy electron's energy will be laterally contained within a radius R_M of the particle's entry axis into the material.

Hadronic Showers. A similar but more complicated phenomenon arises for hadrons as well. High energy hadrons incident upon a material dissipate their energy, in the initial phase, via strong interactions with the atomic nuclei of the material. This creates a number of secondary particles, usually charged or neutral pions or neutrinos, that then continue to decay hadronically, electromagnetically, or escape undetected, respectively. The resulting "hadronic shower" is thus a complicated cascade of different particle species, each dissipating energy through different physics mechanisms (or not at all in the case of neutrinos). Because of this, the energy response of the material will, in general, be different for the different components of the decaying hadron. This makes building a hadronic calorimeter to precisely measure the energy of hadrons a challenge: in general, hadrons will not be measured with energy resolution as good as for EM particles.

In addition, the strong force is much shorter in range than the electromagnetic force. Consequently, the cross section for an incident hadron to interact with a material, expressed in terms of the typical mean free path for a hadronic interaction to occur, is as much as one or two orders of magnitude longer for hadrons than it is for electromagnetic particles. Hadronic calorimeters, therefore, need to be substantially longer to achieve the same energy absorption, or stopping power, as for an electromagnetic calorimeter. The mean free path of a material for hadronic nuclear interactions is parametrized by its nuclear interaction length λ_I.

Ionizing Particles. Electromagnetic particles and hadrons incident on a detector material may not necessarily initiate particle showers as described above. The dominant energy loss mechanism a particle experiences in a material is highly dependent on the mass, energy, and species of the incident particle, as well as the type of detector material, particularly its atomic number.

For instance, the same electron that showers in a suitably designed calorimeter made of a high atomic number, may, in a lower atomic number material, have a negligible bremsstrahlung cross section. Conversely, for the same calorimeter material, an incident, heavier lepton like the muon, even if at the same energy as the electron, may also have a negligible bremsstrahlung cross section.

For a given incident particle type, energy loss mechanisms involving showering generally only turn-on and become dominant at higher particle energies. For certain energy ranges, the incident particle may experience a minimum of energy loss. Typically, for this minimal energy loss regime, the only contributing energy loss mechanism is through atomic ionization (i.e., the liberation of atomic electrons). Known as the minimum ionizing particle (MIP) regime, it is specific to a given particle in a given material. An example energy loss curve is shown in Fig. 2.4 for a muon in copper.

The above considerations factor into the design of tracking detectors, whose primary function is to detect the precise passage point of particle with minimal energy dissipation. For this reason, tracking detector are designed to operate in the MIP regime of the particles they are attempting to track.

2.3 The CMS Detector

The Compact Muon Solenoid (CMS) detector [2] is responsible for collecting the collision data generated by the LHC at the CMS interaction point. This is achieved through a series of concentric cylindrical layers hermetically sealing the central

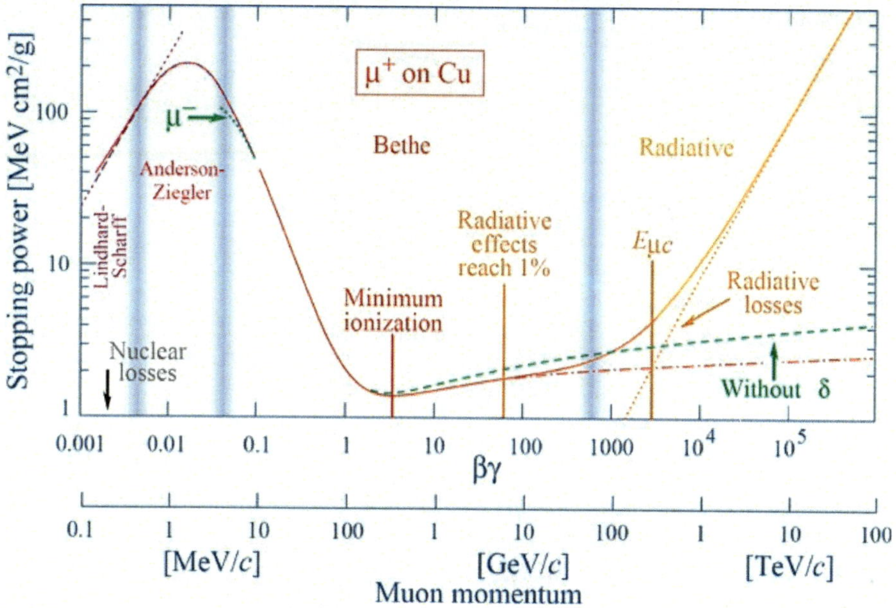

Fig. 2.4 Energy loss quantity and mechanisms for a muon incident on copper at different energies. *Credit* Ref. [1]

collision point. To aid in assembly, the layers are divided into a barrel section, and two circular endcap sections on either side of the barrel section.

The main design driver behind the CMS detector was the CMS solenoid. Relatively compact for its field strength and energy density, the CMS solenoid allowed a single magnet to be used for bending the tracks of light and heavy charged particles alike, a novel concept for its time. The tracking and calorimeter elements are then designed around the dimensions of the solenoid, situated either within or around it. As illustrated in the cutaway view of the CMS detector in Fig. 2.5, the innermost sections comprise the inner tracking system for identifying charged particle tracks. Encasing this is the electromagnetic calorimeter (ECAL) which measures energy deposits from electromagnetic particles, followed by the hadronic calorimeter (HCAL) which measures energy deposits from hadrons. Surrounding the calorimeters is the solenoid generating the magnetic field for the entire detector volume. Outside of the solenoid are the outer tracking detectors, used for identifying the tracks of the heavier muons.

The CMS detector uses a coordinate system centered on the nominal interaction point, at the halfway point through the cylindrical axis of the detector. The axial coordinate is denoted z and the radial distance from this axis r. Angular coordinates are also defined, corresponding to the azimuth (ϕ) and pseudorapidity $\eta = -\ln[\tan(\theta/2)]$, a function of the polar angle θ.

In this section we provide pertinent details about the individual subdetector systems and their general function in the overall data collection strategy. The ECAL is given special focus in the discussion as it is the primary subdetector responsible for

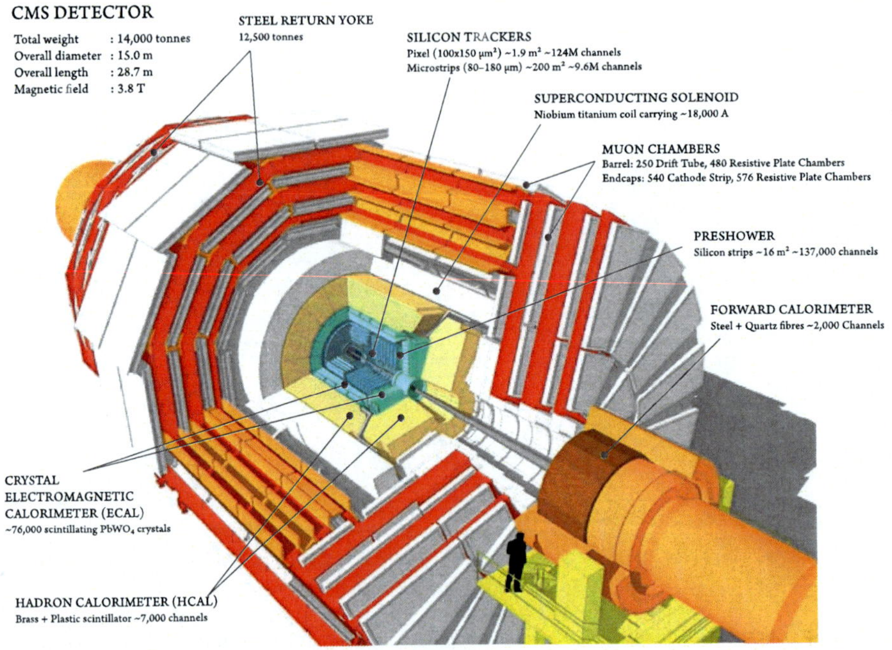

Fig. 2.5 Cutaway view of the CMS detector. *Credit* Ref. [3]

the detection of $\mathrm{H} \to \mathrm{aa} \to 4\gamma$ signatures that are of interest to this physics analysis. A detailed description of all the CMS subsystems can be found in the CMS technical design review [2].

2.3.1 Inner Tracking

The inner tracking subsystems are responsible for recording the position or "hits" of charged particles as they traverse the inner detector volume. As described in Sect. 2.5, sets of tracker hits recorded in the inner tracking system are used to reconstruct full particle trajectories ("tracks") and their point of origin ("impact parameter"). A track may originate from the hard-scatter collision vertex ("primary vertex") or, in the case of a long-lived charged particles, originate some distance away from the primary vertex ("secondary or displaced vertex"). The tracker is designed to enable precise vertex reconstruction in either case, as well good p_{T} determination from the reconstructed track's curvature, as a result of its bending in the CMS solenoid's magnetic field. To meet these requirements, the inner trackers contain the most granular components of the entire CMS detector.

In addition, at distances so close to the interaction point, the inner tracker must be able to tolerate high radiation fluxes, and induce minimal (hadronic) showering, to prevent the creation of secondary particles sprays that would otherwise impact the energy measurement in the calorimeters.

Finally, the tracker must also be designed to stay within allowed trigger rates ($\sim$100 kHz, see Sect. 2.4) given the particle fluxes near the interaction point. With approximately $\sim 10^3$ particles traversing the tracker per bunch crossing, hit rate densities of 1 MHz/ mm^2 at close radial distances ($r \sim 4$ cm), 60 kHz/mm^2 at intermediate distances ($r \sim 22$ cm), and 3 kHz/mm^2 at farther distances ($r \sim 115$ cm) are to be expected.

The inner trackers are thus constructed of thin, extremely granular, silicon sensors that have small radiation and interaction lengths, and intrinsically good radiation hardness. To take advantage of the falling particle flux with radial distance, farther tracking regions are equipped with progressively coarser-grained sensors to reduce the number of read-out channels.

Pixel Tracker. Immediately surrounding the interaction point is the pixel tracker ($3 < r < 16$ cm). It is the most granular subdetector of the inner tracker system—indeed, of the whole of CMS—to enable the most precise determination of the passage of charged particles. The primary goal of the pixel tracker is to provide sufficient hit resolution near the interaction point to enable precise determination of primary and secondary vertices. In particular, precise vertex resolution is critical for correctly identifying which bunch crossing a particle originated from, and for tagging the decay of long-lived b-quark jets ($\sim$mm lifetimes). The pixel tracker is thus constructed of thin, fine-grained, approximately square ("pixel") silicon sensors.

As of the 2016–2017 Phase-1 pixel tracker upgrade [4], the pixel tracker consists of four concentric cylindrical pixel layers covering the barrel region (BPIX) ($|\eta| < 2.5$), and three circular pixel disks (FPIX), on either end of the BPIX, covering the forward range ($1.5 < |\eta| < 2.5$). The current arrangement allows incident charged particles to be sampled at four points throughout most of the covered pseudorapidity. In the BPIX ($3 < r < 16$ cm), the pixels are segmented in the $z - \phi$ plane with a granularity of $\Delta z \times \Delta\phi = 100 \times 150\ \mu\text{m}^2$, and in the FPIX ($|z| = 31$–$51$ cm), segmented in the $r - \phi$ plane with a granularity of $\Delta r \times \Delta\phi = 100 \times 150\ \mu\text{m}^2$. These enable a hit position resolution of between 15 and 20 μm in the barrel layers and about 15 μm in the forward disks. At $|\eta| > 2.5$, pileup contributions dominate, and no attempt is made to record particle tracks.

In total, the current Phase-1 pixel tracker contains 124 million pixels over an area of about 1.9 m^2. The expected hit rate for the BPIX at current LHC instantaneous luminosities (see Sect. 2.1) is between 32 and 580 MHz/cm^2 and between 30 and 260 MHz/cm^2 for the FPIX.

Silicon-strip Tracker. Following the pixel tracker layers is the silicon-strip tracker ($20 < r < 116$ cm). The main task of the silicon strip tracker is to provide adequate track sampling with sufficient hit resolution for track trajectory reconstruction, while staying within the allowed trigger rate. To optimize for these, the silicon-strip tracker is instead composed of several layers of silicon sensors that are only granular along

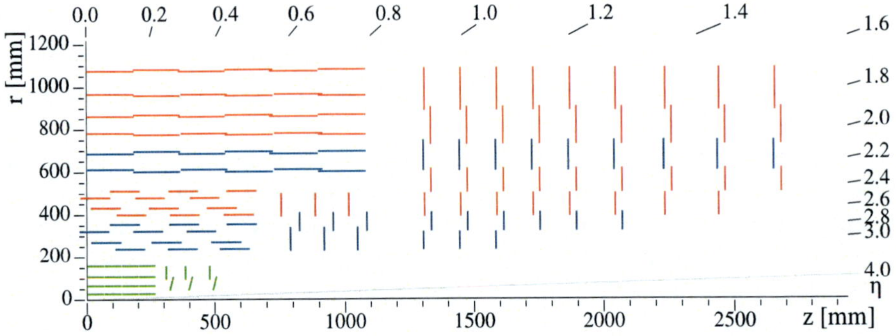

Fig. 2.6 Layout of a longitudinal (r-z) quadrant of the Phase-1 CMS tracking system. The pixel detector is shown in green, while single-sided and double-sided strip modules are depicted as red and blue segments, respectively. *Credit* CMS

the ϕ-direction. Since charged particles only bend in the ϕ direction in the magnetic field of the CMS solenoid, the track p_{T} resolution is driven solely by the curvature in ϕ, and thus only this dimension of the sensor needs to be granular. The silicon-strip tracker, therefore, uses silicon micro-strips that are only finely-segmented in the ϕ-direction ("pitch") in order to sample the passage of charged particles as hits. To reduce the number of strips that need to be read out at each bunch crossing, a tiered strip pitch scheme is used, that grows coarser with radial distance from the interaction point, as the decreasing particle flux allows.

The silicon-strip tracker is further divided into an inner and outer tracking region, as shown in Fig. 2.6. The inner silicon-strip tracker is again composed of cylindrical barrel layers (TIB) and capped off by circular disks (TID). Similarly, the outer silicon-strip tracker is sectioned into cylindrical barrel layers (TOB) and circular endcap disks (TEC).

The TIB-TID subsystem covers the region $20 < r < 55$ cm, $|z| < 118$ cm, and $|\eta| < 2.5$. There are four concentric cylindrical layers in the TIB. They are composed of silicon strips segmented along $z - \phi$ with length $\Delta z = 10$ cm, pitch (width) $\Delta\phi = 80$ μm (layers 1–2) or 120 μm (layers 3–4) layers, and thickness 320 μm. These translate to hit position resolutions of 23 μm and 35 μm, respectively. The TID is composed of four disks on either end of the TIB. They are composed of strips segmented along $r - \phi$ with mean pitch $\Delta\phi = 100$ to 141 μm.

The TOB subsystem covers the region $55 < r < 116$ cm, $|z| < 118$ cm, and $|\eta| < 2.5$. It adds six additional concentric cylindrical layers made of silicon strips segmented along $z - \phi$ with length about $\Delta z = 25$ cm, pitch $\Delta\phi = 183$ μm (layers 1–4) or 122 μm (layers 5–6) layers, and thickness 500 μm. These translate to hit position resolutions of 53 μm and 35 μm, respectively.

The TEC subsystem covers the region $23 < r < 114$ cm, $124 < |z| < 282$ cm, and $|\eta| < 2.5$. It adds nine additional circular disks to each end of the silicon-strip tracker and is made of silicon strips segmented along $r - \phi$ with mean pitch $\Delta\phi = 97$ to 184 μm, and thickness 320 μm (disks 1–4) to 500 μm (disks 5–7).

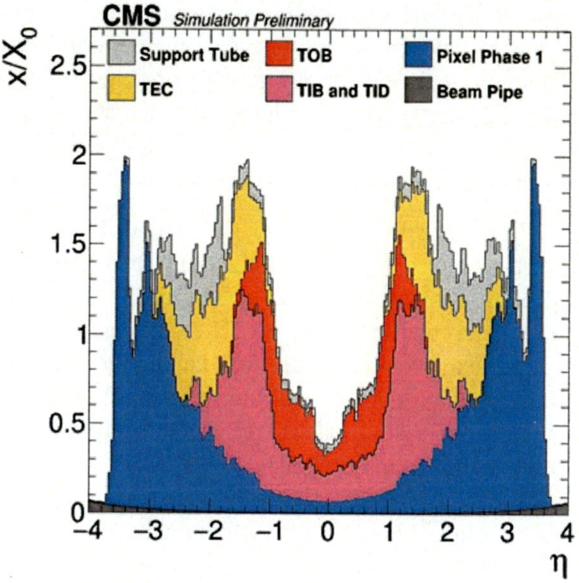

Fig. 2.7 Material budget of the Phase-1 CMS tracking system as a function of η, expressed in the number of radiation lengths X_0. *Credit* CMS

In total, the silicon strip tracker contains 9.3 million strips representing about 198 m^2 of active silicon area. At the specified radial distances and silicon strip pitches, the particle flux at the TIB corresponds to an occupancy rate of 2–3% per strip per bunch crossing, and at the TOB, about 1% per strip per bunch crossing.

For high-p_T tracks ($P_T \sim 100$ GeV), the total inner tracking system achieves a p_T resolution of about 1–2% for $|\eta| \lesssim 1.6$, with a vertex resolution of about 10 μm in the transverse plane (i.e., along the tracker layer). A major source of measurement uncertainty in the reconstructed track p_T resolution is the hadronic showers induced in the tracker support material by the incident particle flux. At high p_T, it accounts for 20–30% of the resolution, while at lower p_T, it is the dominant source of measurement uncertainty. Similarly, for the vertex resolution, scattering from hadronic showers also dominates the measurement uncertainty for low-p_T tracks. At high p_T, however, the vertex resolution is dominated by the position resolution of the first ("seed") pixel hit. Importantly for this analysis, the tracker material also induces electromagnetic showering in high-energy EM particles, before they reach the ECAL. As shown in Fig. 2.7, this varies in η from $0.4X_0$ to $1.8X_0$, with corresponding impacts on energy containment in the ECAL.

2.3.2 *Calorimeters*

The CMS calorimeters are primarily responsible for measuring the energy of electromagnetic particles and hadrons. In contrast to the tracking detectors which seek to measure the position of particle trajectories with as minimal an impact as possible on the particle energy, the calorimeters are designed to fully stop and absorb incident particles in order to measure their energies. A distinct advantage of the CMS solenoid and detector design compared to that of ATLAS is the placement of the calorimeters within the solenoid magnet, at a distance r from the beamline between $129 < r < 286$ cm. This allows the energy of incident particles to be measured more accurately at CMS since the magnet would otherwise act as an absorbing medium.

ECAL. The ECAL is primarily responsible for measuring the energy of incident high-energy electrons and photons. The main goal of the ECAL was to achieve a high-granularity, high-energy precision calorimeter that would enable an accurate determination of the SM $H \rightarrow \gamma\gamma$ mass resonance. It is thus a homogeneous calorimeter, to achieve uniform and predictable energy response, made of a single layer of high-density lead tungstate ($PbWO_4$) crystals that both generate the EM shower and measure its energy. The crystals are scintillators that convert the energy they absorb from the EM shower into scintillation light, in proportion to the incident EM particle's energy. Measuring the light output from the crystals thus allows the EM particle's energy to be determined.

The crystals are grouped into a barrel section (EB), composed of 61200 crystals, that is hermetically sealed off on either end by 7324 crystals grouped into a circular endcap (EE). In the EB, the crystals are arranged in a 170×360 η-ϕ grid and cover the pseudorapidity range $|\eta| < 1.479$. They measure 22 Ã 22 mm^2 at the front face (toward interaction point), or a 0.0174×0.0174 η-ϕ granularity, with a length of 230 mm. In the EE, the crystals are instead arranged in a rectangular x-y grid, covering the range $1.479 < |\eta| < 2.5$. They measure 28.62×28.62 mm^2 at the front face and have a length of 220 mm. Because the EE crystals are arranged in an x-y grid, their effective η-ϕ granularity varies with η, worsening with increasing η. In both the EB and EE, the crystals are tilted toward the interaction point plus some offset ("quasi-projective geometry") to prevent particles from slipping through inter-crystal gaps, as shown in Fig. 2.8. In the EB, the inter-crystal gaps are 0.35 mm, except between supporting structures (submodules, containing 400–500 crystals), where they are instead 0.5 mm.

The $PbWO_4$material is chosen primarily for its high density, allowing for a short radiation length ($X_0 = 0.89$ cm) and Moliere radius ($R_M = 2.2$ cm) that enables a compact, high-granularity ECAL to be built. Within the length of the crystals, they accommodate a total radiation length of $25X_0$ ($24.7X_0$) in the EB (EE). Similarly, more than 90% of the energy of an incident EM particle will be laterally contained in a 3×3 cluster of crystals. The crystals are also engineered for radiation hardness to tolerate the LHC beam environment, with scintillation decay times of order the LHC bunch crossing spacing (25 ns).

However, due to the relatively low light yield of $PbWO_4$crystals, amplifying photodetectors must be used to collect the light output from the crystals. In the EB, these are accomplished using avalanches photodiodes (APDs). In the EE, where the magnetic field lines are bent and higher radiation levels are present, vaccuum phototriodes (VPTs), a type of photomultiplier, are instead used.

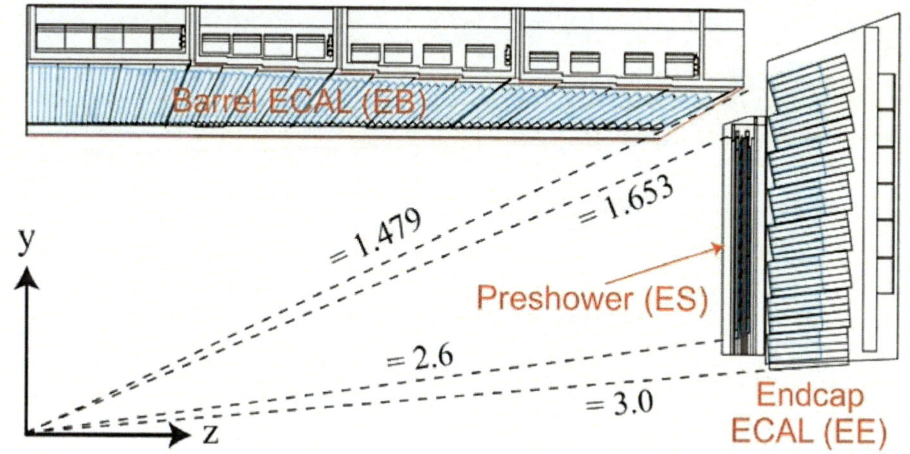

Fig. 2.8 Layout of a longitudinal (r-z) quadrant of the ECAL. *Credit* Ref. [5]

The signals from the photodetectors are then passed to the electronics for processing. The electronics are separated into on- and off-detector electronics that optimize between bandwidth and radiation constraints. The on-detector electronics, optimized for bandwidth, consist of a multi-gain pre-amplifier to further amplify the signal from the photodetector, followed by an analog-to-digital converter (ADC) to digitize the signal into a pulse shape of 10 amplitudes ("digitized hit"), separated by 25 ns each (i.e., the bunch crossing spacing). Signals from different crystals are then grouped by the Front-End (FE) card (5×5 crystals in the EB, variable in the EE). The FE group of crystals also defines a trigger tower (TT), used to construct "trigger primitives". The trigger primitives are coarser sums of the underlying crystal signals that are quickly sent to the Level-1 trigger at the full LHC clock rate (40 MHz), to determine whether the event is worth keeping. Only if a decision from the global Level-1 trigger to keep the event is received ("L1-accept") are the individual crystal signals read out to the off-detector electronics. Among other things, the off-detector electronics are responsible for collating all the data from the different FEs, validating them, and building a complete ECAL event description.

Even with the triggering system (described in more detail in Sect. 2.4), the ECAL electronics (and, indeed, the rest of the CMS data acquisition chain) would be unable to manage the full read out of all ECAL crystals in the triggered event. The off-detector electronics thus implement a selective read-out processor (SRP) for selecting regions-of-interest around energetic deposits. The SRP makes these decisions at the trigger-tower (TT) level, determining whether to read out the crystals underlying the TT based on the energy sum of the TT (E_{TT}). The SRP has two thresholds: if the E_{TT} is above the highest threshold, all the crystals in the TT plus the crystals of all immediately adjacent TTs (in EB, 3×3 TTs or 15×15 crystals) are fully read-out. If the E_{TT} is between the two thresholds, only the crystals for that TT are read-out. If the E_{TT} falls below the lowest threshold, only crystals with an energy above $3\sigma_{\mathrm{noise}}$

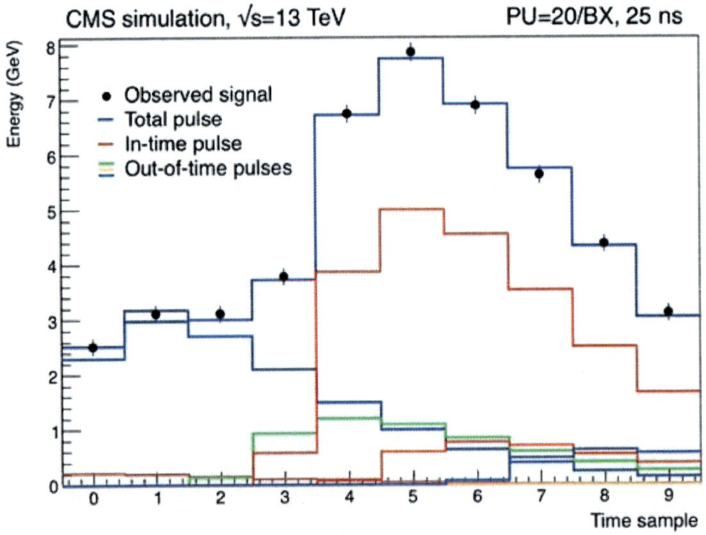

Fig. 2.9 The ECAL multifit applied to the total observed signal pulse yields a fit for the in-time pulse contribution, plus up to nine additional out-of-time pulses. *Credit* CMS

are read out from the TT, so-called "zero-suppressed" read-out. Importantly, this means energetic deposits appear as rectangular blocks of non-zero depositions in the data. During offline reconstruction, a filter is applied to only keep data from crystal deposits above a certain threshold.

At this stage, the crystal deposits are still in the form of digitized amplitudes (10 per crystal per event). In order to reconstruct an energy measurement from the crystal, the multifit "hit reconstruction" algorithm is used [6]. For each digitized pulse, the multifit algorithm fits for upto 10 pulse shape templates, one for each 25 ns time step, as shown in Fig. 2.9. This allows the hit reconstruction to separate the contribution from the "in-time" pulse that triggered the event, from those of adjacent bunch crossings due to pileup. The "reconstructed hit" energy is then determined from the peak of the fitted in-time pulse shape, after converting from ADC counts to GeV.

In order to function in its role as a high-precision calorimeter, the ECAL energy response must be calibrated in both space and time to achieve per mille accuracy. The dominant source of absolute energy variation comes from optical transparency variations in the $PbWO_4$crystals due to the annealing and recovery of the crystals under irradiation from the LHC beam. The transparency loss is measured by pulsing a laser light onto each photodetector, in sequence, every ≈ 89 μs, at the end of the LHC beam cycle, or the so-called "abort gap". A full scan of the detector is completed in about 20 mins. This is then used to determine an appropriate correction for the energy response to scintillation light in each crystal. A finer correction of the absolute energy scale variations in time is achieved by exploiting the position of the $Z \rightarrow e^+e^-$ resonance. Sources of relative energy variation are due to differences in relative response between crystals in the detector as a result of irradiation differences but are also due to manufacturing variances. To correct these, appropriate inter-crystal calibration are derived by exploiting the $\pi^0 \to \gamma\gamma$ and $\eta \to \gamma\gamma$ resonances in low-energy, EM-enriched jets ($E \sim 1\text{GeV}$) and the ϕ-symmetric distribution of electrons in $W \to e\nu$ decays.

From test beam studies with electrons with energies between 20 and 250 GeV, the ECAL energy resolution was measured to be

$$\frac{\sigma}{E} = \frac{2.8\%}{\sqrt{E/\text{GeV}}} \oplus \frac{12\%}{E/\text{GeV}} \oplus 0.3\%, \tag{2.4}$$

consistent with estimates from simulation. For energies $E \approx 60\text{GeV}$, or about the median energy of a $\text{H} \to \gamma\gamma$ photon, the energy resolution is about 0.5%. The typical electronics noise ("pedestal") in an EB (EE) crystal is measured to be 40 (150) MeV.

Lastly, installed in front of each EE face (as viewed from the interaction point) are the preshower subdetectors (ES). The ES is a sampling calorimeter, composed of two pairs of alternating layers of lead absorber and silicon sensors. These cover the pseudorapidity range $1.653 < |\eta| < 2.6$. The ES is considerably finer-grained than the ECAL, with each silicon layer composed of strips with an active pitch of 61×1.9 mm, with the second layer oriented perpendicular to the first. Originally envisioned to aid in the discrimination of neutral pions (e.g., $\pi^0 \to \gamma\gamma$) from single photons, unfortunately, due to the spurious signals generated by the large neutral pion flux from hadronic interactions in the tracker, the full granularity of the ES has yet to actually be exploited. Moreover, due to its lead absorber plates, the ES contributes $3X_0$ of material budget preceding the EE, impacting energy resolution in the EE quite significantly relative to that in the EB. We speculate on the possibility of using the ES to augment the reconstruction of $a \to \gamma\gamma$ decays in the endcap in Sect. 10.

HCAL. The HCAL is responsible for measuring the energy of hadrons typically contained within jets. As noted in Sect. 2.2, a major constraint in constructing hadronic calorimeters is the much longer nuclear interaction lengths required for the hadronic shower to develop and dissipate its energy (as opposed to electromagnetic showers), requiring correspondingly deeper hadronic calorimeters. Therefore, to maximize the nuclear interaction lengths that can be contained within the physically allowed space for the HCAL, as is common for hadronic calorimeters, the HCAL is designed as a sampling calorimeter that alternates between high-density absorber layers optimized for generating hadronic showers, and sensitive layers for sampling the energy output of the hadronic shower.

The HCAL contains a barrel section (HB) covering the pseudorapidity range $|\eta| < 1.3$ with 5.82–10.6λ_I (increasing with η), sealed off on either end with an endcap section covering $1.3 < |\eta| < 3$ with about $10\lambda_I$. This is further augmented by an additional outer barrel section (HO), lying just outside the CMS solenoid, for a total of $> 10\lambda_I$ (including the solenoid) at $|\eta| < 1.3$. The ECAL ($|\eta| < 3$) contributes about $1\lambda_I$ preceding the HB-HE. An additional, forward endcap (HE) covers the range $3 < |\eta| < 5$, however, this is only designed to be sensitive to the

EM component of hadronic showers. For $|\eta| \gtrsim 2.5$, pileup contributions primarily dominate the hadronic deposits.

In the HB-HE sections, the absorber layers are made from brass, and the sensitive material are made from plastic scintillators. In all, there are upto 16 layers of absorber-scintillator pairs, depending on η. The layers are segmented into η-ϕ towers. For $|\eta| < 1.6$, the granularity is $\Delta\eta \times \Delta\phi = 0.087 \times 0.087$, or about the coverage of 5×5 EB crystals, while for $|\eta| > 1.6$, the ϕ granularity is a coarser $\Delta\phi = 0.174$. The scintillation signals are collected by means of wavelength-shifting (WLS) fibres embedded in the plastic scintillators, which are then routed to a hybrid photodiode (HPD) just outside of the scintillator for light measurement. Before being read out to the electronics for processing, the signals from the (upto) 16 layers are grouped into up to four depth segments, with the signals from the underlying layers summed over. Thus, the effective depth segmentation Δd of the HB-HE is only $\Delta d \lesssim 4$.

Together with the ECAL, the ECAL+HCAL energy resolution was measured in pion test beams to be

$$\frac{\sigma}{E} = \frac{2.8\%}{\sqrt{E/\text{GeV}}} \oplus \frac{12\%}{E/\text{GeV}} \oplus 0.3\%, \tag{2.5}$$

consistent with estimates from simulation. As explained in Sect. 2.2, the energy resolution for hadronic showers is significantly worse than for electromagnetic-only showers in the ECAL. The typical electronics noise in an HB-HE tower is measured to be about 200 MeV.

2.3.3 *Magnet System*

As suggested by the name of the experiment, the primary consideration in the design of the CMS magnet system was the development of a relatively compact solenoid with the bending power needed to deflect and precisely measure the momentum of high energy muons. This is achieved through a single, high-field 3.8 T superconducting solenoid magnet with a steel return yoke structure. At a diameter of 6 m and length of 12.5 m, describing the CMS solenoid as "compact" may hardly seem apt, however, for the field energy it stores ($E \sim 2.6$ GJ), it has one of the highest energy storage capacities per unit mass ($E/M \sim 12$ kJ/kg). For context, the ATLAS detector has three magnet systems: an inner solenoid, a barrel toroid, and two endcap toroids. The ATLAS magnet with the largest energy storage of the three, the barrel toroid, has only half the energy storage capacity ($E \sim 1$ GJ) and one-sixth the energy per unit mass ($E/m \sim 12$ kJ/kg) of the CMS solenoid, but is almost thrice the diameter and twice the length! By comparison, the CMS solenoid is indeed quite compact.

Solenoid. The inner bore of the CMS solenoid encases all of the inner tracking systems and both calorimeters. It is composed of a 4-layer winding of NbTi, stabilized and reinforced with aluminum and aluminum alloy, respectively. It has a radial thick-

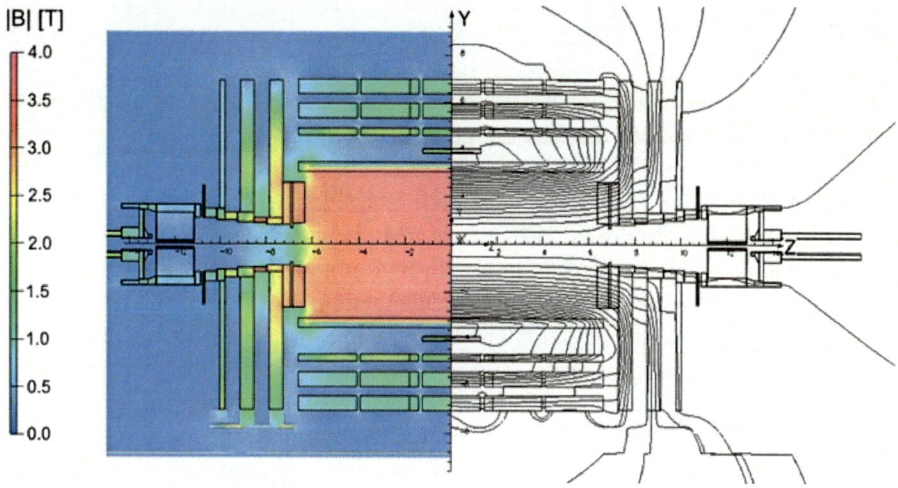

Fig. 2.10 Mapping of the CMS solenoid magnetic field strength (left half) and field lines (right half) as measured from the deflections of cosmic ray muons. *Credit* Ref. [7]

ness of 630 mm and cold mass of 220 tons. The NbTi has a superconducting critical temperature of $T_c = 7.3$ K, at the peak field strength of 4.6 T. During operation, the solenoid is cooled using liquid helium to a temperature of $T = 4.5$ K, generating a field of 3.8 T at a nominal current of 19 kA. To maintain its temperature, the solenoid is encased in a vacuum cryostat.

A detailed description of the magnetic field lines of the solenoid throughout the CMS detector volume is obtained by a complement of direct measurements performed during the original installation of the detector, and indirect measurements from the deflection of cosmic ray muons [7]. The uncertainty in the measured field strength is between 2 and 8%, and in agreement with numerical simulations to within 5 mT. As shown in Fig. 2.10, inside the solenoid volume, particularly in the region of the inner trackers, the magnetic field is effectively uniform so that charged particle tracks can be approximated as helices. Outside and toward the opening of the solenoid, however, the magnetic field lines are highly non-trivial, which must be taken into account when reconstructing the tracks of muons in the outer tracking system.

Yoke. The return yoke immediately surrounds the CMS solenoid, extending to a radius of 14 m, and into which the outer tracking systems are embedded. It is a steel structure composed of a barrel section split into five slices or "wheels", and two endcap sections split into three disks each, capping off either end of the solenoid. The barrel and endcap structures do not have a rounded cylindrical shape but rather have a 12-sided cross section, giving the CMS detector its distinctive, red, dodecagon shape. These flat sectors are designed to accommodate alternating layers of muon detection chambers and steel return yoke, with three layers in total, as described in the following sections.

2.3.4 Outer Tracking

The muon or outer tracking subsystems are responsible for recording the position or hits of muons as they traverse the outer detector volume. Muons are long-lived, charged leptons, much larger in mass than electrons ($\sim 200m_e$), and are often indicative of interesting high energy physics processes, e.g., $\mathrm{H} \rightarrow \mathrm{ZZ}^* 4\mu$. Because of their heavier mass, they do not radiate intense EM showers in the ECAL like electrons do, instead only interacting minimally through ionization, and otherwise escaping through the ECAL. However, due to the magnetic field of the CMS solenoid, they have curved trajectories. In particular, as the muons cross the solenoid, the direction of the magnetic field flips, in turn, causing the muons to bend in the opposite direction. Muons thus have a distinct reconstructed track signature that few other background objects can fake, making them a valuable handle for analysis. The goal, then, of the muon trackers is to identify the passage of muons, both for the purposes of triggering and offline analysis, and to provide sufficient hit sampling and precision to be able to measure their track p_T. These are achieved by interleaving tracking layers into the steel return yoke outside the CMS solenoid: since only muons are likely to survive this far, as the direction of the magnetic field flips outside the solenoid, muons are identified by a reversing track curvature, from which their p_T can also be measured.

There are three muon tracking subsystems: the drift tubes (DT) in the barrel, the cathode strip chambers (CSC) in the endcap, and the resistive plate chambers (RPC) in both the barrel and endcap. All three utilize the ionizing effect of charged muons traversing a gas-filled volume to register a track hit. While the DT and CSC combined have sufficient timing and position resolution to efficiently identify and associate muons with the correct bunch crossing, due the historical uncertainty in background particle flux, the RPCs were envisioned to provide an unambiguous redundancy through their superior timing resolution.

DT. Surrounding the CMS solenoid is the DT muon tracking system. It is a cylindrical barrel section only covering about the same length as the CMS solenoid, and the pseudorapidity range $|\eta| < 1.2$.

The DT has a deeply heirarchical layer structure. At the highest level, the DT is divided into four concentric cylindrical layer groups or "stations" interleaved within the three layers of the steel return yoke. Within each station, there are eight layers of gas "chambers". In each chamber, there are then twelve (in stations 1–3) or eight (in station 4) aluminum layers in groups of four, called superlayers (SLs). Each aluminum layer is, in turn, composed of series of parallel (cathode) drift tubes running the length of the chamber, and filled with a gas mixture of 85% Ar + 15% CO_2. Running inside of each drift tube is an (anode) wire for collecting the electrons from the gas as it is ionized by passing muons. In the innermost (first four aluminum layers) and outermost (last four aluminum layers) SL of each chamber, the drift cell/wires run parallel to the beamline (z-direction) providing fine segmentation along the ϕ direction. For the chambers with three SLs (in stations 1–3), the middle SL (middle four aluminum layers) runs in the ϕ-direction to provide fine segmentation along z-direction.

In total, there are about 172,000 drift tubes. Each drift tube has a cross section 42 mm wide and 13 mm thick. With an electron drift velocity of about 54 μm/ns, this translates to a time response of 380 ± 4 ns, or a hit position resolution of about 170 μm.

CSC. The CSC is an endcap-only section to complement the barrel-only coverage of the DTs. It is the circular endcap section capping off the barrel section of the total CMS volume on either side, and provides muon tracking for the pseudorapidity range $0.9 < |\eta| < 2.4$, partially overlapping with the DTs. Because of the higher background particle flux and the non-uniform magnetic field lines in the forward regions of CMS solenoid, the CSC is optimized separately from the DTs which experience little background flux and a mostly uniform magnetic field.

In each endcap, the CSC is broken down into four layer groups or stations, interspersed within the steel flux return plates. Each station is divided radially into one to three annuli, and azimuthally into 36 or 72 overlapping sectors. Each division contains a trapezoidal chamber spanning an arc of 10° or 20° and length corresponding to the width of the annulus. These are multiwire proportional chambers composed of seven (cathode) layers (in z-direction) or "panels" sandwiching six gap layers 7 mm thick filled with a gas mixture of 40% Ar + 50% CO_2 + 10% CF_4. The cathode panels are split into strips running the radial length of the chamber and segmented in ϕ, and hence the name cathode strip chambers. The gas gaps, on the other hand, contain a plane of anode wires that run in the ϕ-direction (perpendicular to the cathode strips), spaced apart by 3.2 mm in the radial direction. The orthogonal orientation of the cathode strips from the anode wires allow for position determination in the full r-ϕ plane.

In total, the CSC contains about 5000 m^2 worth of cathode strip area, > 50 m^3 of gas volume, and 2 million anode wires. Each CSC chamber has a comparable time response to a DT drift tube of around $380 \pm \lesssim 5$ ns, and position resolution of 80 μm.

RPC. The RPC complements the DT and CSC muon subsystems by providing superior timing resolution of about 1 ns, to provide unambiguous triggering of the correct bunch crossing, but with coarser position resolution. It contains cylindrical barrel layers (RB) as well as circular endcap layers (RE) on either end. Due to budgetary constraints, the endcap layers only cover the pseudorapidity range $|\eta| < 1.6$.

The RB layers, six in total, are divided among the same four stations. In stations 1–2, there two RB layers, one before and after each DT layer. In station 3–4, there is one RB layer each, before the DT layer. The redundancy in stations 1–2 ensures even low-p_T tracks that do not reach the later stations are sufficiently sampled. Each RB layer is broken axially into five segments ("wheels"), and azimuthally into twelve sectors of flat "double-gap modules". The RB layers thus share the same distinctive dodecagon (12-sided) cross-section as the CMS return yoke. Each double-gap module is made of a pair of parallel plate capacitors that share a common, instrumented area in between them. Each parallel plate capacitor is filled with a gas mixture of 96.2% $C_2H_2F_4$ + 3.5% iC_4H_10 + 0.3% SF_6, and hence the name resistive plate chambers. The inner plates of the double-gap modules sandwich a plane of up to 96 sensitive strips, finely-segmented in the ϕ-direction, that collect the ionization charge induced by the passing muon to provide position measurement.

The RE layers, three in total, are located in stations 1–3, before (station 2) or behind (stations 1 and 3) the CSC layers. Each circular layer is divided radially into three annuli, and azimuthally into 36 (outer two annuli) or 72 (innermost annulus) sectors. Each trapezoidal division likewise contains a double-gap module of gas-filled parallel plate capacitor pairs sharing a common sensitive area of up to 32 strips, finely-segmented in the ϕ direction.

In the RB, the double-gap modules cover a total area of 2400 m^2 with 80,640 strips.

2.4 The CMS Trigger System

At a collision rate of 40 MHz, the LHC proton-proton bunches generate data at a rate far too high than could be stored during the time of the LHC's conception, decades before "Big Data" solutions would become mainstream. Moreover, the overwhelming majority of these collisions—by several orders of magnitude—consists of soft-scattering QCD events (c.f. Fig. 2.2) lacking the energy to create the kind of high energy, hard-scattering events that are of interest to CMS. For these reasons, CMS employs an event filtering or *triggering* scheme, to restrict data collection to only the most interesting events. It is the first phase of event selection that must be factored into any physics analysis.

The trigger selection is performed in two stages that progressively reduce the event rate using progressively more nuanced selection criteria. The first stage, the Level 1 (L1) trigger, is performed by high-speed electronics using coarser grained detector information. It reduces the event rate from 40 MHz to about 100 kHz. This is followed by the second stage, the high-level trigger (HLT). The HLT, in contrast, is software-based and uses the full detector granularity. While slower, the HLT can be much more precisely and flexibly tuned to select high interest events. The HLT reduces the L1 event rate down to less than 1 kHz. Only events passing the HLT are stored for physics analysis.

While a number of detector signatures are used to define high interest events, a general feature of high energy processes is the presence of tracks or calorimeter deposits with large transverse momentum $P_T \gtrsim 10$ GeV. As a matter of convention, because the calorimeters measure energy not momentum directly,[4] the transverse projection of the energy deposited in the calorimeters, E_T, is used. Due to the large computational cost of inner track reconstruction, the L1 trigger only uses information from the calorimeters and the muon systems to make decisions about which events to keep. The HLT uses information from all subdetectors, including the inner trackers.

In the following subsections, we give a brief overview of the data reconstruction architecture of the triggers.

---

[4] At these energies, the distinction is negligible and, in practice, the terms are used interchangeably.

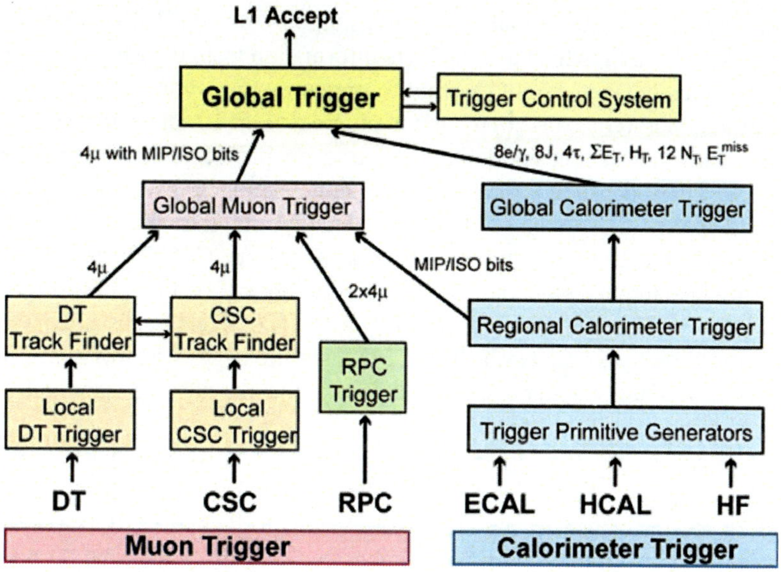

Fig. 2.11 Data reconstruction pipeline of the CMS L1 trigger system. *Credit* Ref. [2]

2.4.1 L1 Trigger

The L1 Trigger is the very first stage of event filtering applied to collision events. Because of the immense speed at which the L1 needs to make decisions, hardware-based technology is primarily used, in the form of programmable FPGAs or ASICs. For the components of the L1 Trigger system that are situated on-detector, the hardware are additionally designed to be radiation tolerant.

The L1 Trigger data reconstruction pipeline is organized into a hierarchy of local, regional and global components, as illustrated in Fig. 2.11. At the level closest to the individual subdetectors are the local triggers, also known as the Trigger Primitive Generators (TPG). These collect the E_{T} deposits from the ECAL or the HCAL trigger towers ($\Delta\eta \times \Delta\phi \approx 0.087 \times 0.087$), and the hits from the muon trackers. In addition, the TPGs are responsible for assigning the trigger data to the correct bunch crossing.

On the calorimeter pipeline, the Calorimeter Trigger, the ECAL and HCAL trigger primitives from the TPG are then forwarded to the Regional Calorimeter Trigger (RCT). The RCT further sums these TPG trigger towers into RCT towers of about 4×4 TPG trigger towers. The RCT then uses this information to construct crude electromagnetic particle candidates, and also overall RCT tower E_{T} sums to be used for the next step of the calorimeter trigger. The RCT also calculates information about MIPs, relevant for muon reconstruction, which is instead passed to the Muon Trigger chain. The next step from the RCT is the Global Calorimeter Trigger (GCT). The GCT uses information from the RCT to construct jet candidates, missing E_{T} candidates, and a more nuanced classification of electromagnetic candidates into isolated or non-isolated. Finally, the information from the GCT is passed to the Global Trigger (GT), which also collects information from the Muon Trigger chain, to make a final decision about whether to trigger on the event.

A similar hierarchy exists in the muon pipeline of the L1 chain, the Muon Trigger. At the local level, the DT and CSC muon trackers provide either track hits or track segments. These are then passed to the Regional Muon Trigger (RMT) which performs complete track reconstruction (for the outer, muon tracking region only) and subsequently constructs muon candidates for the DT and CSC. The muon candidates from the RMT are then passed to Global Muon Trigger (GMT). The RPC, because of its superior timing resolution, is able to reconstruct complete tracks and passes muon candidates directly to the GMT. The GMT uses the trigger information it receives from the three muon tracker to collate and refine the muon trigger candidates, before finally passing these to the GT.

At the end of the L1 trigger pipeline is the GT. Using the particle candidate information it receives GCT and the GMT, the GT takes the decision of whether to accept or reject an event at the L1. If the event is to be kept, a "L1-accept" (L1A) is passed back down to the individual subdetectors. During the time it takes the L1 pipeline to reach a decision, the full granularity detector data is held in memory within each subdetector's on-detector electronics. If an L1A signal is received from the GT, the full data are subsequently transferred to the HLT.

2.4.2 HLT Trigger

The HLT is the second and final stage of event filtering applied within the CMS Trigger system. As the HLT takes as input the reduced L1 rate, it can afford to trade some speed for accuracy. The HLT thus uses software-based decision making using the full detector granularity—including the inner tracker information—to reduce the event rate to $\lesssim 1$ kHz. These are processed on-site, although above ground, by a CPU farm of more than 10,000 cores. Only events selected by the HLT are recorded for physics analysis.

The physics candidates reconstructed by the HLT, however, are not directly used by physics analyses, and are only used for taking decisions about whether to keep or reject events. Instead, the stored events selected by the HLT are forwarded to a separate pipeline for "offline" (i.e., post-collection) physics reconstruction. Both HLT and offline reconstruction use similar software-based algorithms. A detailed description of the physics reconstruction algorithm, therefore, is deferred to the following Sect. 2.5 on offline physics reconstruction. The main difference, because the HLT must still make time-critical decisions, is that the HLT runs some of these algorithms at reduced accuracy, specifically those relating to track reconstruction, and with a cutoff in the maximum number of reconstructed candidates.

In addition to filtering events, the HLT also categorizes events based on topology. The HLT decisions are thus broken down into various topology categories or "menus", each catering to the identification of the signatures of a particular kind or class of signal topologies of interest. For instance, the menu used for this analysis (see Sect. 6) is that for events containing two energetic, isolated photons with invariant mass above 90 GeV. In addition to its obvious relevance to the SM $H \to \gamma\gamma$ search, it is also widely used for searches involving energetic diphotons such those predicted by a number of BSM scenarios, and of course, for the case of merged photons in $H \to aa \to 4\gamma$.

The same event may fall under more than one menu, and analyses may choose to chain together multiple menus to achieve the highest signal efficiency and/or background rejection. As a matter of accounting, the final data sets stored for physics analyses are categorized based on menu groups, with again the possibility of events occurring in multiple data sets. While the creation of HLT menus are flexible, and can change within any period of data-taking, the primary driving constraint in the evolution of HLT menus is that they fall within the total allowed rate budget of the HLT. As the total rate varies with the supplied LHC instantaneous luminosity, it is not uncommon to see the thresholds in these menus changed from year to year, in step with changes in the beam luminosity.

In the following Section, we go into detail about how events selected by the HLT are processed by the full offline physics reconstruction algorithm to generate the particle candidates that ultimately find use in most physics analyses.

2.5 Physics Object Reconstruction

While the detector-level data are the fundamental representations of the CMS experimental data, the typical CMS physics analysis is performed using *physics objects*. These are the reduced, idealized, particle-like representations expressed in terms of particle species and four-momenta. While a key strategy of this analysis is using machine learning methods directly on detector data (see Sect. 4), the physics object convention is nonetheless needed to define the analysis phase space in a manner that is compatible with wider CMS conventions. In particular, the event selection and data control regions are, at least in part, defined in terms of reconstructed physics object-derived quantities, to maintain orthogonality with complementary analyses.

Physics object reconstruction is performed by the CMS Particle Flow (PF) algorithm [8], a primarily rule-based algorithm for reconstructing particle properties given the various signals provided by the different CMS subdetectors. To begin with, the PF algorithm takes the signals collected from the different subdetectors and uses these to build basic PF elements, namely, tracks and calorimeter clusters. A linking process is then performed to match track elements with calorimeter clusters to reconstruct the different identifiable physics analysis objects in the event. An illustration of how the different subdetector elements are used to identify the different particle types is shown in Fig. 2.12. Specific particle species classifications are not always attempted, except where feasible. Instead, particle candidates are grouped into object categories: photons, electrons, muons, QCD jets, and missing momentum.

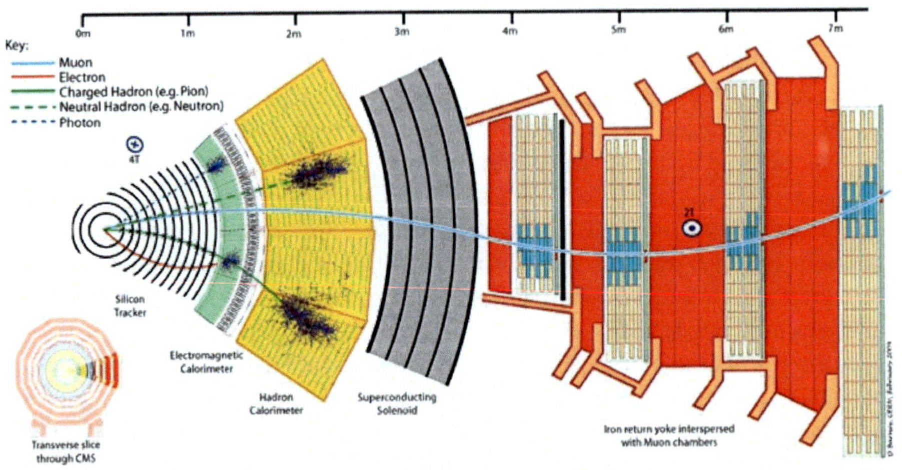

Fig. 2.12 Illustration of how different CMS subsystems are used to identify different particle species. *Credit* CMS

In the subsections below, we describe the construction of the basic PF elements from detector inputs, followed by the construction of PF objects or candidates from the PF elements.

2.5.1 PF Elements

The PF elements constitute the basic building blocks of physics objects. In this subsection, we introduce these elements and how they are derived from the detector inputs.

Tracks and Vertices. Track and vertex reconstruction is relevant for identifying and measuring the momentum of charged particles, and for identifying the displaced decay vertices associated with b-quark jets. The track reconstruction takes as inputs the hits deposited in the tracking layers, fits particle trajectories through these, then uses the fit parameters to extract or calculate various quantities of interest.

The tracks are built in several stages using an iterative, combinatorial, Kalman filter algorithm. The basic strategy of iterative tracking is to identify the "easiest" tracks first (e.g. large p_{T} or produced near the interaction point), remove the hits associated with these from the collection, then repeat the process. At each iteration, the process starts by identifying candidate "seed" hits in the pixel tracking layers. Initial track candidates are then formed by fitting curves through the seeds that maximize the track fit. Using a Kalman filter, these initial track candidates are then extrapolated to the layers of the silicon-strip tracker. Compatible hits along the path of the track are subsequently incorporated into the track fit and the fit recalculated. After exhausting all track candidates, complete tracks passing certain fit criteria are kept. Upto five of the best-fitting tracks are then excluded from the collection of track hits used for the next iteration. Upto six iterations are performed. The HLT and offline particle reconstruction differ primarily in how many iterations are allowed for track reconstruction. At the HLT, only two iterations are performed.

From the fitted track, the impact parameter is extracted to give the position of the track's vertex. In the inner trackers where the magnetic field of the CMS solenoid is, to good approximation, uniform, the tracks are described by helical sections. By measuring the curvature of the fitted track ("sagitta"), the momentum of the charged particle associated with the track can thus be extracted. For muon tracks in the outer trackers, however, the magnetic field is highly irregular. A computationally derived model of the magnetic field is instead used [9], through which the fitted track is propagated to extract the muon momentum.

Calorimeter Clusters. Calorimeter clustering is relevant for measuring the four-momenta of electromagnetic particles (photons and electrons) and stable, neutral hadrons in the case of the ECAL, and of charged hadrons in the case of the HCAL. In conjunction with the reconstructed track information, cluster information is also used to distinguish charged particles from neutral ones.

Clustering is performed separately for each calorimeter subsystem: EB, EE, or ES for ECAL, and HB or HE for the HCAL. Calorimeter clusters are formed by first identifying cluster seeds, or calorimeter cells that represent a local maxiumum above some energy threshold. Clusters are then grown from the seeds by incorporating neighboring cells that share at least a corner in common with a cell already associated with the cluster, with energy above twice the noise threshold.

For clustering in the ECAL specifically, superclusters or groups of closely spaced clusters are also defined. Since photons are able to convert to e^+e^- pairs, and electrons may emit a bremsstrahlung photon, it is desirable to have such arrangements still categorized as a single physics object. The ECAL supercluster, therefore, is formed by admitting additional cells from an extended window in ϕ for some narrow spread in η. While beneficial in most physics analyses, the ECAL supercluster algorithm leads to a host of challenges with the a $\rightarrow \gamma\gamma$ topology. In particular, the reconstructed momentum can be underestimated (see Sect. 4.2), the hits of the softer photon from the a can be dropped (see Sect. A.2), and the number of physics objects the a $\rightarrow \gamma\gamma$ is reconstructed as can become ambiguous (see Sect. 6.4).

Finally, to account for energy losses due to the energy thresholds in the clustering, showering in the preceding tracker material, and leakage through calorimeter gaps, the cluster (or supercluster, in the case of ECAL) energy is calibrated as a function of energy and η.

2.5.2 PF Objects

Given the basic PF elements described above, the PF algorithm links together the different elements to enable coherent identification and reconstruction of the various

PF objects used in the typical physics analysis. For this thesis, we note that only photon—and to a lesser degree, electron—candidates are used. Descriptions of the other physics objects, however, are still presented for completeness.

Photons. An isolated photon candidate is seeded from an ECAL supercluster with transverse energy $E_T > 10\,\text{GeV}$, with no associated track reconstructed in the inner tracker. Additionally, the photon candidates is required to be isolated from other nearby tracks and calorimeter clusters, and not have an associated HCAL deposit containing more than 10% of its ECAL supercluster energy.

The reconstructed energy of the photon candidate is determined from the supercluster energy, after applying energy corrections. The photon direction is determined from the barycenter of its associated supercluster.

Electrons. Isolated electron candidates are instead seeded by tracks with an associated ECAL supercluster. Similar isolation requirements as with the photon are also imposed. Additional requirements, however, are imposed on the quality of the track and its compatibility with the ECAL supercluster.

The reconstructed energy of the electron is obtained from a combination of both the corrected ECAL energy and the momentum of the associated track. The electron direction is chosen to be that of the associated track.

Muons. Isolated muon candidates are identified by the presence of reconstructed tracks in the outer tracking system that satisfy compatibility requirements with a matching track from the inner tracking system. Any additional tracks and calorimeter deposits within an angular cone of $\Delta R = \sqrt{\Delta\phi^2 + \Delta\eta^2} < 0.3$ of the identified muon track are required to sum to a p_T not exceeding 10% of the muon track p_T.

For $P_T < 200\,\text{GeV}$, the muon momentum and direction are given by the associated inner track's momentum and direction. Above this threshold, they are determined based on the combination of inner and outer track that gives the best track fit.

Jets. The jet candidates are those that remain after isolated photons, electrons, and muons have been identified and removed from the pool of particle elements. These are further classified as charged hadrons (e.g., $\pi^\pm$, $K^\pm$, protons), neutral hadrons (e.g., K^0_L or neutrons), nonisolated photons (e.g., π^0 that have failed the isolation criteria for photon candidates), or, more rarely, nonisolated muons (e.g., from the early decays of charged hadrons). We note, however, that in most analyses that study jets, these further classification are not commonly used. In particular, for this analyses, we do not directly use jets categorized as nonisolated photons. Nonetheless, the isolation of, for instance, π^0 in jets is a continuous spectrum. In practice, a number of these will fall within even the isolated photon category (see Sect. 8.2).

Any ECAL (HCAL) clusters not linked to a track are identified as nonisolated photons (neutral hadrons). Beyond the tracker acceptance ($|\eta| > 2.5$), neutral hadrons can no longer be distinguished from charged hadrons. The energy of the nonisolated photons (neutral hadrons) is determined from the underlying, corrected ECAL (HCAL) cluster energy (or ECAL+HCAL cluster energy, in the case $|\eta| > 2.5$).

The remaining HCAL clusters are then linked to one or more matching tracks, together with any matching ECAL clusters (potentially matched to a track each), to form charged hadrons. The energy of the charged hadron is either the sum of the calorimeter cluster energies, or the sum of the matched track momenta, whichever is larger. The difference between the two can be used to further categorize particles in the jet.

While not discussed further here, it is worth pointing out that the particles identified in the jet candidate may alternatively be clustered using dedicated jet clustering algorithms. Indeed, for most CMS analyses studying jets, jet candidates clustered using the anti-k_{T} algorithm are by far more widely used than the PF jets described above.

Missing Transverse Momentum. To accommodate the possibility of particle decays that do not interact with the detector, e.g., neutrinos, a final particle category known as missing transverse momentum, $\mathbf{p}_{\mathrm{T}}^{\mathrm{miss}}$, is introduced.[5] Since the LHC proton-proton beams have no net momentum in the transverse plane, momentum conservation in the transverse plane can be invoked to infer the presence non-interacting "invisible" decays from the momentum vector needed to balance the visible decays:

$$\mathbf{p}_{\mathrm{T}}^{\mathrm{miss}} = - \sum_{i=1}^{N_{\mathrm{particles}}} \mathbf{p}_{\mathrm{T}}, i, \tag{2.6}$$

where $\{\mathbf{p}_{\mathrm{T}}, i\}_{i=1}^{N_{\mathrm{particles}}}$ is the set of transverse momentum vectors of the visible decays reconstructed by PF. In practice, however, jet energy mis-measurements and the presence of pileup and underlying event can influence the measurement $p\mathrm{T}^{\mathrm{miss}}$. For analyses seeking invisible decays, therefore, additional corrections and optimization are often applied to improve the measurement of $\mathbf{p}_{\mathrm{T}}^{\mathrm{miss}}$.

2.6 Detector Simulation

A principal feature of the end-to-end particle reconstruction technique used in this analysis for $\mathrm{a} \rightarrow \gamma\gamma$ mass reconstruction, is the use of (simulated) detector-level data—not high-level reconstructed physics objects (see Sect. 7)—for training. In this section, therefore, we present a overview of the elements represented in the CMS detector simulation, with a particular focus on the ECAL, and how these have been validated against data. These are intended to complement the dedicated data versus simulation comparisons presented in Sect. 7 that are specific to the $\mathrm{a} \rightarrow \gamma\gamma$ application.

For its detector simulation, the CMS experiment uses a *Geant4*-based simulation toolkit [10]. Using Monte Carlo methods, the *Geant4* toolkit [11] provides a platform for the simulation of the passage of particles through matter. As relevant for the

---

[5] As noted in Sect. 2.4, since the calorimeters measure energy and not momentum, as a matter of convention, the missing transverse momentum is also known as the missing transverse energy $E\!\!\!/_{\mathrm{T}}$, even though, strictly speaking, energy is a scalar quantity.

ECAL, the detector simulation consists of three main parts: the detector description, the tracking of particles passage through the detector material and the subsequent detector response, and the modeling of the electronics readout [12].

The detector description includes the measurements, alignments, and material properties of the ECAL crystals themselves but also materials involved with structural support, cooling, readout electronics, and cabling. These are based on the detailed, as-built blueprints of the CMS detector, as well as actual measurements performed prior to detector assembly, in the case of material weights.

The tracking of the simulated particles through the detector material accounts for the influence of the magnetic field of the CMS solenoid. The scintillation of the ECAL crystals in response to incident particles, however, is modeled simplistically. The response is parametrized in terms of an effective conversion factor from the deposited hit energy to the mean number of photoelectrons generated by the interaction. One conversion factor is determined for the EB and EE separately, and accounts for the excess noise generated by the avalanche process in the photodiodes, and the non-uniform scintillation generated along the crystal's length—that is, that the signal pulse generated in the photodiodes has a sharp rise but extended decay tail—including any potential energy leakage behind the crystal, if not fully contained.

The electronics readout is modeled by emulating the digitization of the photodiode signal pulse through the multi-gain pre-amplifier plus analog-to-digital converter to output a 10-step digitized signal pulse (see Sect. 2.3.2).

Finally, energy calibrations are applied to the individual, simulated ECAL crystals based on the average transparency measured for each crystal in a given year of data-taking. To improve the simulation fidelity, a crystal cluster-based containment correction is applied to account for the sensitivity of the electromagnetic shower response to the exact position at which an incident particle enters the crystal face.

The ECAL detector simulation has been validated using both test beam and collision data. Test beam data using electron beams have shown the transverse electromagnetic shower shapes to be within 1% of simulation, and for energy resolution comparisons, within a few percent for most of the energy range [12]. Collision data were used to validate the electromagnetic shower widths in the η-direction, which were also shown to be in good agreement with simulation [10].

A more general description of the CMS detector simulation for other subdetectors can be found in [10].

References

1. Dorigo T, Kieseler J, Layer L, Strong G (2020) Muon energy measurement from radiative losses in a calorimeter for a collider detector. Tech Rep
2. Chatrchyan S et al (2008) The CMS experiment at the CERN LHC. JINST 3:S08004
3. Sakuma T, McCauley T (2014) Detector and event visualization with SketchUp at the CMS experiment. J Phys Conf Ser 513:022032
4. T. T. G. of the CMS Collaboration (2020). The CMS phase-1 pixel detector upgrade

5. Benaglia A (2014) The CMS ECAL performance with examples. J Instrum 9:C02008–C02008 Feb
6. Sirunyan AM et al (2020) Reconstruction of signal amplitudes in the CMS electromagnetic calorimeter in the presence of overlapping proton-proton interactions. J Instrum 15:P10002–P10002 Oct
7. Chatrchyan S et al (2009) Precise mapping of the magnetic field in the CMS barrel yoke using cosmic rays. JINST 5:T03021, 35
8. Sirunyan AM et al (2017) Particle-flow reconstruction and global event description with the CMS detector. JINST 12(10):P10003
9. Klyukhin V (2021) Design and description of the CMS magnetic system model. Symmetry 13(6)
10. Banerjee S (2012) CMS simulation software. J Phys Conf Ser 396:022003
11. Agostinelli S et al (2003) GEANT4: a simulation toolkit. Nucl Instrum Method A506:250–303
12. Rovelli C (2008) Validation of the simulation of the CMS electromagnetic calorimeter using data. Tech Rep. CERN, Geneva

Chapter 3
Theory and Phenomenology

In this chapter, we provide a basic theoretical basis for the physics search performed in this analysis. In Sect. 3.1, we introduce the standard model of particle physics (SM), currently the best available theory for the interaction of the known fundamental particles. A basic description of the fundamental symmetries and mechanisms involved in the SM is provided. We follow this with a more focused discussion of the different physics sectors of the SM, with an emphasis on aspects relevant to the search performed in this analysis. Then, in Sect. 3.2 we highlight some of the challenges associated with the SM that motivate the need for a deeper physical theory. In Sect. 3.3, we introduce the extended Higgs sector that is the focus of the beyond the SM (BSM) physics search in this analysis, and lay out its theoretical and experimental motivations. The chapter concludes with a discussion of the phenomenology of the particular $\mathrm{H} \rightarrow \mathrm{aa} \rightarrow 4\gamma$ signal process chosen for analysis in Sect. 3.3.1.

3.1 The Standard Model

The Standard Model of particle physics describes the dynamics of the known elementary particles, interpreted as quantum fields, and their interactions with one another. It accounts for three of the four known fundamental interactions or forces: electromagnetism, the weak force, and the strong force. As of this writing, it is the leading theory of fundamental particle interactions and, save for a few recent tensions with experiment, has been consistent with most laboratory tests of predicted particle properties.

The dynamics of elementary particles are described by the SM in terms of the interaction of quantum *gauge field* theories. In gauge field theory, particles are represented as quantum fields whose dynamics are encoded in a quantity called a Lagrangian density, or simply, a Lagrangian $\mathcal{L}$. The process of extracting particle dynamics from the Lagrangian, which can then be compared against experimental observables, is a follows. From the most general form of the Lagrangian, one requires the participating quantum fields to obey symmetries—under which the Lagrangian is unchanged—to

M. Andrews, *Search for Exotic Higgs Boson Decays to Merged Diphotons*,
Springer Theses, https://doi.org/10.1007/978-3-031-25091-0_3

constrain the form of the Lagrangian. To test a particular interaction, a perturbative expansion is performed[1] in powers of the coupling strength between the interacting fields. Each order in the expansion represents a series of individual interactions and typically visualized as a Feynman diagram. For a given initial and final state of particles, one then sums all relevant individual processes in the expansion, up to the desired order of precision, to obtain a prediction for the associated observables, e.g., the total cross section times branch fraction for the initial states to decay into the final states.

In the SM, the symmetries imposed on the particle fields that are found to be consistent with experiment are those of global Poïncare symmetry, and of local symmetry under the unitary group $SU(3) \times SU(2) \times U(1)$. The Poïncare symmetry is the symmetry under the Lorentz transformations of Special Relativity, plus translations in space-time. Requiring particle fields to obey Poincare symmetry begets the spin representation of the elementary particles: scalar bosons (spin-0), fermions (spin-$\frac{1}{2}$), and vector bosons (spin-1)[2].

On the other hand, the $SU(3) \times SU(2) \times U(1)$ symmetry corresponds to the conservation of color charge under interactions with the gauge field G, weak isospin charge under the gauge field Y, and weak hypercharge under the gauge field B, respectively. After spontaneous symmetry breaking, as explained below, the gauge fields undergo mixing to manifest themselves as the known physical bosons: the gluons, the W and Z bosons, and the photon, along with the interactions they mediate That is, by requiring the quantum fields to be gauge-invariant in addition to obeying Poincare symmetry, one recovers the known subatomic interactions: quantum chromodynamics (QCD) for the strong force, the theory of weak interactions for the weak force, and quantum electrodynamics (QED) for the electromagnetic force, respectively.

The need for spontaneous symmetry breaking arises because the original gauge fields (G, Y, B) are all forbidden from "naively" carrying mass terms in their Lagrangian, as these would violate gauge invariance. Historically, gauge field theory was motivated by successes in QCD and QED where the gauge bosons were massless, as observed experimentally and as required by gauge invariance. However, the W and Z bosons, conceived to be the gauge bosons of a new interaction, were decidedly massive, at odds with the requirements of gauge invariance.

This conflict is resolved by introducing the Higgs mechanism: by adding a complex scalar field, the Higgs field, with a potential, the Higgs potential, whose energy minimum does not coincide with the zero-value of the Higgs field, a so-called non-zero vacuum expectation value (VEV), the resulting interactions with the gauge fields result in terms in the Lagrangian that mimic gauge boson masses. Indeed, interactions of the Higgs field with the fermionic fields generate masses for the fermions as well. As a quantum field itself, the Higgs field is associated with a new particle, the Higgs boson.

---

[1] If working in a regime where a perturbative expansion is valid, otherwise non-perturbative methods must be brought to bear.

[2] The putative graviton would be the only spin-2 particle.

While the exact shape of the Higgs potential as a function of the Higgs field is determined by a number of free parameters, the most general renormalizable construction satisfying non-zero VEV, is the so-called "Mexican hat" potential. In this potential, gauge symmetry is manifest at the zero-value of the Higgs field but not at the (usually stable) minima of the Higgs potential. At energy regimes below the Higgs VEV ($E \approx 250\,\text{GeV}$), as the state of the universe cools toward a stable energy minimum of the Higgs potential, the gauge symmetry is said to be *spontaneously broken*.

In order to generate masses for the W and Z bosons of experiment, the weak interaction has to be unified with QED into an abstract electroweak (EW) gauge group $\text{SU}(2) \times \text{U}(1)$, that is spontaneously broken into the sole $\text{U}(1)_{\text{QED}}$ symmetry of QED. In particular, for the neutral Z boson to acquire a mass while leaving the photon massless, the $\text{U}(1)$ gauge group prior to symmetry breaking must differ from the $\text{U}(1)_{\text{QED}}$ after symmetry breaking, so that the Y and B states are allowed to mix. Under this scheme, the Higgs field must be a doublet of the $\text{SU}(2)$ group and, after symmetry breaking, the Y and B gauge bosons mix to give the massive W and Z bosons, and the single massless photon of experiment. The gluons, being massless gauge bosons, are unaffected by the Higgs mechanism.

While color (from $\text{SU}(3)$) and electric charge (from the remaining $\text{U}(1)_{\text{QED}}$) are conserved after spontaneous symmetry breaking, weak isospin and weak hypercharge, in general, are no longer conserved under physical interactions. The original QCD gauge group can then be trivially combined with the EW gauge group to form the full SM gauge group of $\text{SU}(3) \times \text{SU}(2) \times \text{U}(1)$.

To summarize, the SM Lagrangian $\mathcal{L}_{\text{SM}}$, before spontaneous symmetry breaking, can be concisely expressed as:

$$\mathcal{L}_{\text{SM}} = \mathcal{L}_{\text{gauge}} + \mathcal{L}_{\text{EW}} + \mathcal{L}_{\text{QCD}} + \mathcal{L}_{\text{Higgs}} \tag{3.1}$$

where the $\mathcal{L}_{\text{gauge}}$ contains the "free field" gauge terms, $\mathcal{L}_{\text{EW}}$ describes electroweak interactions, $\mathcal{L}_{\text{QCD}}$ describes strong interactions, and $\mathcal{L}_{\text{Higgs}}$ contains the Higgs field and potential, and describes the interaction of the Higgs field with the gauge and fermion fields, imparting mass upon them. A table of the known elementary particles and their basic properties is given in Fig. 3.1.

While the full gauge symmetries of the SM are manifest prior to spontaneous symmetry breaking, it is worthwhile to discuss some aspects of the SM after spontaneous symmetry breaking as this represents the energy regime of all of LHC physics, and consequently, this analysis. At this regime, QED, QCD, weak interactions, and the Higgs sector are approximately decoupled, and can be described separately, as we do next.

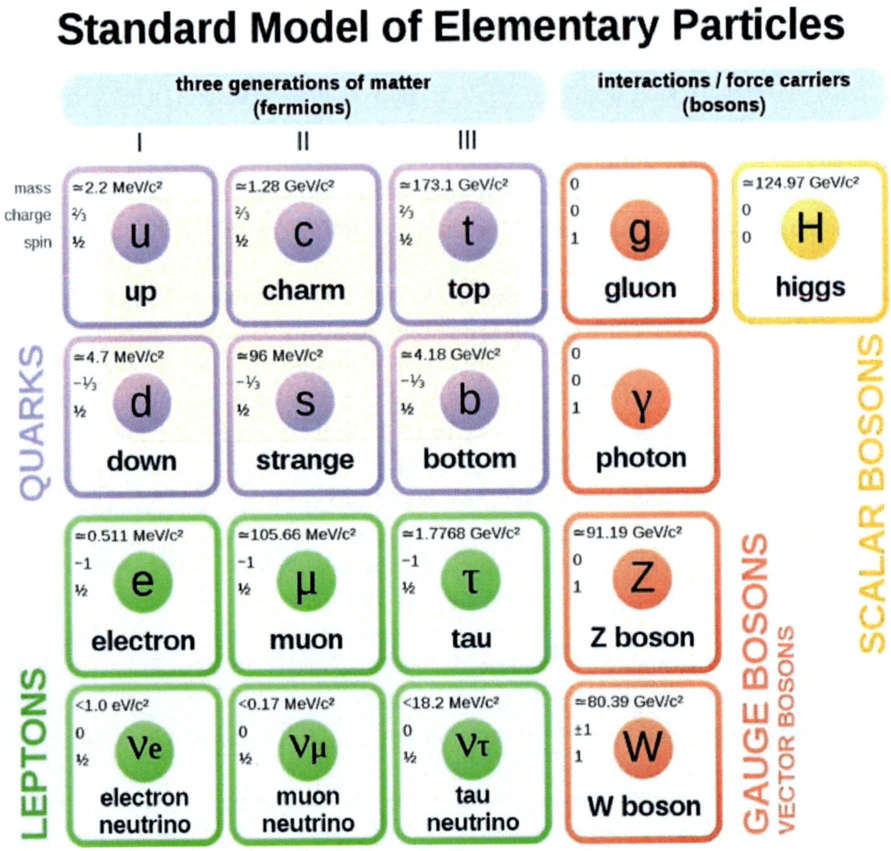

Fig. 3.1 The particles of the standard modelCredit: Wikipedia

3.1.1 QED

The QED component of the SM describes how electrically charged particles interact with photons. Since the electromagnetic gauge group of QED is preserved after spontaneous symmetry breaking, electric charge is conserved under QED interactions. The basic QED interaction vertex is given in Fig. 3.2. By chaining together the basic QED vertex, more complex interactions can arise. A few important examples of these that feature in this analysis are the processes of electron bremsstrahlung and electron-positron pair production, arising as a result of a high energy electron or photon, respectively, interacting with the charged nucleus of some (detector) material, via QED. Representative Feynman diagrams are shown in Fig. 3.3. In the case of electron bremsstrahlung (Fig. 3.3a), the incident electron is converted to an electron plus photon, while for pair production (Fig. 3.3b), the incident photon is converted to an electron plus positron. These are, for instance, the main processes governing the detection of electromagnetic particles in calorimeters (see Sect. 2.2).

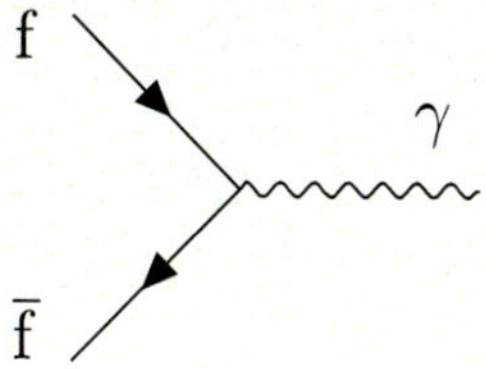

Fig. 3.2 QED interaction vertex between a photon (γ) and a fermion/antifermion (f/$\overline{f}$)

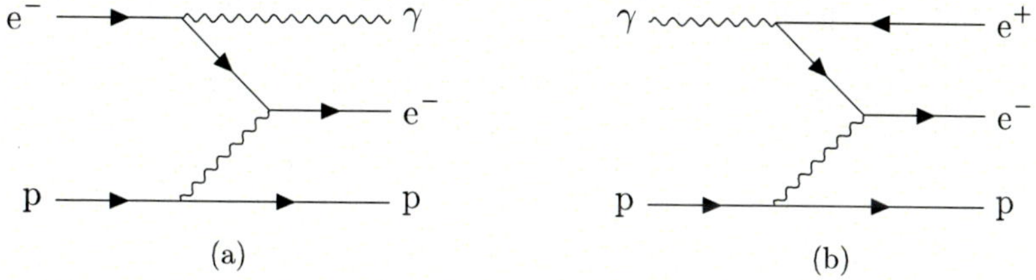

Fig. 3.3 QED nuclear interactions between a proton (p) and an electron/positron (e^+/e^-) or photon (γ): (3.3a) electron bremsstrahlung, (3.3b) e^+e^- pair production

While QED was originally formulated for electrically charged fermions like the charged leptons and the quarks, in EW theory, after spontaneous symmetry breaking, the photon is also allowed to interact with the electrically charged $W^\pm$ bosons (see Sect. 3.1.3).

3.1.2 *QCD*

The QCD component of the SM describes how color-charged particles, of which quarks are the only known example, interact with gluons. QCD has several properties that distinguish it from the simpler physics of QED. While color charge is conserved under QCD interaction, because of the larger SU(3) symmetry group to which QCD belongs, QCD contains three charge quantum numbers, denoted red, green, and blue, in contrast to the single electric charge of QED. In addition, unlike QED, QCD is a non-Abelian gauge theory, i.e., the generators of its group do not commute. A consequence of this is that the gauge boson of QCD, the gluon, is color charged. In fact, there are eight gluons, based on the allowed color combinations of the group. Gluons may therefore directly engage in color self-interactions. The interaction vertices of QCD are thus as given in Fig. 3.4. The first (Fig. 3.4a) describes the interaction of quarks with gluons, while the latter two (Fig. 3.4b, c) describe 3- or 4-point gluon self-interaction.

A rich family of quark bound states arises from the more complex color structure of QCD. There are the mesons, which are quark-antiquark pairs bound in color-anticolor, and the baryons, which are three quarks bound in all three color charges.

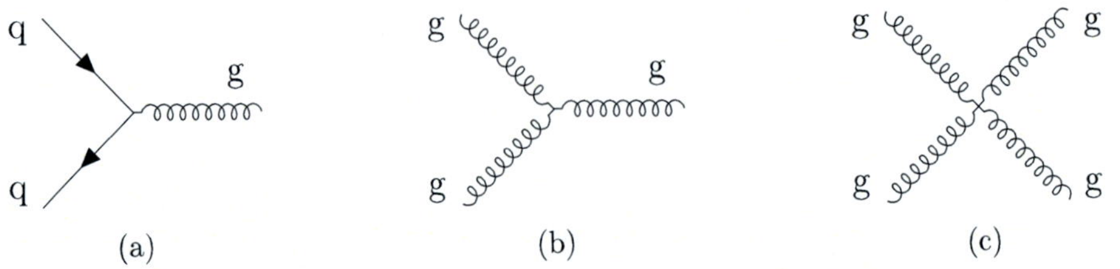

Fig. 3.4 QCD interactions vertices between quarks (q) and gluons (g): (3.4a) qg interaction vertex, (3.4b) 3-point gluon self-interaction, (3.4c) 4-point gluon self-interaction

By analogy to the mixing of the colors of light, both mesons and baryons, collectively known as hadrons, are "colorless" states.

That gluons are allowed to directly self-interact leads to another important feature of QCD, namely, that the QCD coupling strength between quarks *increases* with distance. Thus, when attempting to separate bound quarks, say a $q\bar{q}$ meson, it becomes energetically favorable for the meson to create a new $q\bar{q}$ pair from the vacuum, resulting in the formation of two $q\bar{q}$ mesons instead. Because of this, bare quarks are not observed in nature, a phenomenon known as confinement, resulting in only colorless QCD states being observed in nature.

Confinement carries important phenomenological consequences for high energy colliders: the creation of high energy quarks or gluons invariably results in a chain of hadrons being produced instead. The resulting stream of collimated hadrons is known as a hadronic jet or simply a jet, and the process by which it is formed is called jet fragmentation or hadronization. Hadronic jets are an important signature of high energy phenomena at particle detectors like CMS. At lower energies, lacking the energy needed to collimate into a jet, a diffuse spray of hadrons is instead produced. Such soft scattering interactions, for instance, between glancing proton collisions at the LHC, while not indicative of high energy phenomena, form a dominant component of the background processes observed at hadron colliders like the LHC. The top quark, because of its large mass, has a lifetime shorter than the hadronization timescale, and is thus notably exempt from hadronization. Instead, the top quark decays via the weak force to a W boson plus a quark. It is therefore one of the only ways in the SM to study a "bare" quark.

Confinement leads to a host of computational challenges as well. Below the confinement scale ($\sim 10^2$ MeV), i.e., when trying to separate quarks, that the QCD coupling strength increases with distances means that a perturbative expansion in orders of the coupling strength no longer converges. Known as the non-perturbative QCD regime, no rigorous tools yet exist for predicting the interactions of quarks at this regime, and heuristic tools must be relied on instead. Above the confinement scale, however, the running of the QCD coupling strength implies that the interaction strength between quarks becomes small. Thus, at high energy regimes, the quarks behave as effectively free particles, a complementary phenomenon known as asymptotic freedom. This regime is manifest in the primary hard scattering interactions that take place at high energy colliders like the LHC, and for which perturbation theory remains valid.

3.1.3 *Weak Interactions*

The weak interaction component of the SM describes how particles charged under weak isospin interact with the physical (i.e., post symmetry-breaking) weak gauge bosons, namely, the electrically charged $W^{\pm}$ bosons, and the electrically neutral Z boson. Since weak interactions derive from the weak SU(2) isospin gauge group, which is broken by the Higgs mechanism, weak isospin charge, denoted up or down, is not conserved under weak interactions.

Since SU(2) is also a non-Abelian theory, the weak gauge bosons are also charged under weak isospin—as the gluons were under color charge—and thus self-interact as well. However, the EW boson state mixing induced in the original $SU(2) \times U(1)$ by spontaneous symmetry breaking introduces a number of unique consequences not seen for the other non-Abelian theory of QCD, but are required for agreement with experiment.

First, the mixing of EW boson states allows the physical EW bosons to not only self-interact, but to interact directly with the other EW bosons as well.

Second, just as the fermions needed to carry electric or color charge in order to interact with the photon and gluon, respectively, so too do the fermions need to carry weak isospin charge in order to interact with the weak bosons. An important distinction of weak interactions, however, is that the W boson only interacts with left-hand chirality fermion particles and right-handed chirality fermion antiparticles. Only these fermions carry weak isospin charge as a result. The weak interaction thus contains the only processes in the SM able to violate parity and charge-parity conservation, as first observed in experiment.

Third, the Z boson is able to couple to all charged fermions, of any chirality. Thus, while the Z boson is itself not electrically charged, under weak interactions, it is able to couple to electrically charged fermions nonetheless. The available weak interaction vertices after spontaneous symmetry breaking are shown in Fig. 3.5.

The first (Fig. 3.5a) and second (Fig. 3.5b) vertices are the analogs of the fermion vertices of QED (Fig. 3.2) and QCD (Fig. 3.4a). The first vertex is the so-called charged current (CC) weak interaction, allowing an up-type fermion to decay into down-type one, or vice-versa, if kinematically allowed. Of historical note, before the idea of weak isospin—which is conserved under weak interactions—was known, the charge of weak interactions was thought to be flavour (see Fig. 3.1), which is distinctly not conserved under CC interactions. The CC interaction was thus also acknowledged as the only SM process not to conserve flavor.

The second vertex (Fig. 3.5b) is the neutral current (NC) of the weak interaction which does conserve flavor. While not manifest in the basic interaction vertex, so-called flavor changing neutral currents (FCNC) appear to be suppressed by loop contributions, though searches for FCNC processes continue to be an important test of the SM. Of note, the NC mediates a process that features prominently in various calibration tasks at CMS, namely, the decay of the Z boson to an electron-positron pair, $Z \to e^+e^-$. Since few other processes in the SM produce two isolated electrons

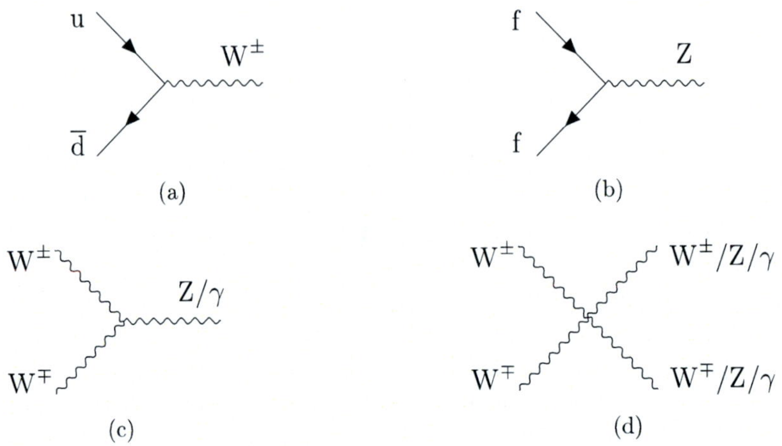

Fig. 3.5 Weak interaction vertices of EW bosons ($W^{\pm}$, Z, γ): (3.5a) charged current interaction with left-handed up-type fermion (u) and right-handed down-type antifermion ($\bar{d}$), (3.5b) neutral current interaction with fermions (f), (3.5c) 3-point gauge coupling, and (3.5d) 4-point gauge coupling

near the Z boson mass resonance, such decays provide a rare opportunity to study electromagnetic particles with high precision.

The third (Fig. 3.5c) and fourth (Fig. 3.5d) vertices are the 3-point and 4-point gauge coupling interactions, respectively. They are the analogs of the corresponding gluon self-interaction vertices (see Fig. 3.4b, c) but, because of the mixing from spontaneous symmetry breaking, occur between the different EW gauge bosons, as noted earlier.

3.1.4 *Higgs Sector*

The Higgs sector describes the interactions of the Higgs boson resulting from the Higgs mechanism. While the Higgs mechanism was motivated by the need to impart mass upon the weak gauge bosons, it generates a mechanism for imparting mass upon the fermions as well, via the Yukawa coupling. The Higgs boson thus directly couples to all massive particles in the SM. In addition, the Higgs boson is itself a massive boson. While not a gauge boson, the Higgs boson obtains a mechanism for self-interaction via the Higgs potential. The Higgs interaction vertices are thus as given in Fig. 3.6.

The first (Fig. 3.6a) describes the interaction of the Higgs boson with the massive gauge bosons. The second (Fig. 3.6b) describes the Yukawa coupling of the Higgs boson to the (massive) fermions. The third (Fig. 3.6c) and fourth (Fig. 3.6d) describe the 3- and 4-point self-interaction of the Higgs boson.

While the Higgs boson does not directly couple to the massless gauge bosons, it is able to do so indirectly via loop contributions. An important example of this is the

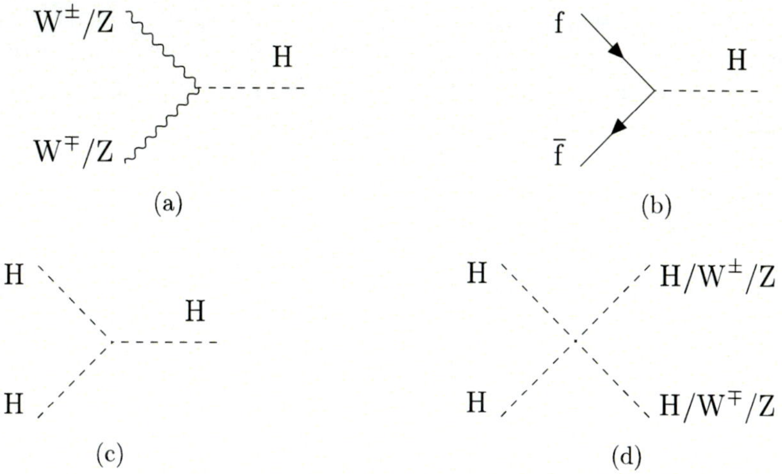

Fig. 3.6 Higgs boson (H) interaction vertices: (3.6a) with EW gauge bosons ($W^{\pm}$, Z), (3.6b) Yukawa coupling with fermions (f), (3.6c) 3-point Higgs self-interaction, (3.6d) 4-point Higgs self-interaction or with EW gauge bosons

Fig. 3.7 The $H \rightarrow \gamma\gamma$ decay of the SM

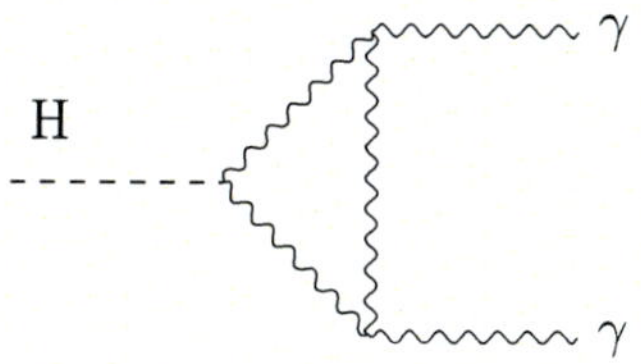

decay of the Higgs boson to two photons, one of the major background processes involved in this analysis. The dominant contribution to this process comes from the W-boson loop, as shown in Fig. 3.7.

3.2 Challenges with the SM

As successful as the SM has been in agreeing with laboratory—typically, collider-based—experiments, it is in disagreement with many astronomical observations, at times, glaringly so [1].

Dark Matter. The SM is unable to explain the overwhelming majority of the energy content of the universe, estimated to be over 90% in the form of dark matter and dark energy. Dark matter is thought to be responsible for the gravitiontal stability of galaxies, and the structure of the universe at-large. It is hypothesized to be some form of matter that interacts gravitationally but not electromagnetically, i.e., does not emit light. While the exact nature of dark matter is not known, no candidate in the SM exists that fits the required properties. A number of BSM models, including those involving supersymmetry or axion production, have historically gained favor for predicting particles with properties similar to that expected of dark matter. Dark energy, on the other hand, is an even bigger mystery still, existing only as the unknown force of expansion in the universe at-large, against the attraction of gravity.

Baryon asymmetry. Even within the scope of known, baryonic matter does the SM fail to give a proper accounting of astronomical data. As far can be determined, all astronomical matter is of the particle type, as opposed to the antiparticle type. Yet assuming the universe was created with equal parts of each, no mechanism exists in the SM for initiating a departure from these conditions, or for generating the large asymmetries between matter and antimatter decays. For the latter, the SM, through processes that do not respect charge-parity symmetry, provides for the preferential decay to matter particles in the decays of strange- and charm-quark mesons. However, by running these through models of the formation of the early universe, these CP-violating processes are found to be insufficient to reproduce the matter distribution present in the current universe. This is the so-called baryon asymmetry problem. As of this writing, however, early hints have begun to show of tensions with SM predictions in B-meson decays [2].

Strong CP problem. As noted earlier, while the SM allows for CP-violating processes, as has been observed experimentally for EW processes, not all of these provisions have been realized, particularly in the QCD sector. In addition to furthering the baryon asymmetry problem, this is also indicative of a fine-tuning issue, where the value of certain free parameters in the Lagrangian seem to be coincidentally aligned to give cancellation of CP-violating process in QCD. While more a "naturalness" issue than an explicit disagreement between theory and experiment, the apparent fine-tuning is thought to be due to the action of some as-yet unknown mechanism. One potential solution that has been proposed is the existence of an axion particle. In Pecci-Quinn theory [3], the CP-violating phase of QCD is treated, as the Higgs field was to boson mass, as yet another quantum scalar field, which, post-spontaneous symmetry breaking, settles to a value close to zero. As was the case for the Higgs field, this additional scalar field also results in a new particle state, identified as the axion. Moreover, to achieve resolution of the strong CP problem, the axion is required to be a CP-odd pseudoscalar [4]. Current constraints on axion production, however, limit the viability of the simplest axion models in solving the strong CP problem.

Parameter fine-tuning. More generally, the SM Lagrangian is rife with various occurrences of free parameter fine-tuning, particularly those relating to particle mass hierarchies. For instance, the GeV mass scale of the Higgs entails the cancellation of loop contributions to the Higgs mass that are of order the Planck scale (10^{19} GeV), to multiple orders of the perturbative expansion. There is no known mechanism in the SM for generating cancellations across such disparate mass scales over several orders. Indeed, this was one of the prime motivations for the introduction of the superpartner particles of supersymmetry. More generally, BSM models attempt to address this divide in mass scale hierarchies by re-casting the SM coupling strengths as dynamic couplings that vary with energy, which, at around the Planck scale, unify into a single interaction. Such theories are known as grand unified theories (GUTs).

More recently, a few laboratory-based experiments have managed to demonstrate departures from SM predictions: the non-zero mass of neutrinos, for instance, or at least hints of potential departure: the early measurements for K^* decays to electrons and muons indicating tension with lepton flavor universality, and a potential deviation in the muon magnetic dipole moment.

Lastly, in the quest to unify the sub-atomic forces with gravity into a "theory of everything", a quantum description of gravity is still absent in the SM. In all sub-atomic regimes accessible by experiment, gravitational forces are all but vanishingly small. While this has certainly not prevented the theory community from speculating how such a theory would look, the ability to produce "microscopic" blackholes would likely be needed to test any of these.

Part of the challenge in addressing the above problems is the unknown energy regime at which deviations from the SM are thought to occur. The production of dark matter, for instance, may only occur at energies much higher than can be accessed at the 10^4 GeV scale of the LHC. After all, the early universe was exposed to energy densities much higher than can be produced at any laboratory to date. Indeed, GUTs predict unification to occur at 10^{15}–10^{16} GeV, well-beyond even the wildest imaginings the accelerator community has on their drawing boards.

In this analysis, therefore, we take we take a more modest and model-agnostic approach in attempting to resolve any of the above challenges with the SM. The last few years have seen many of the high energy physics community's most favored theories slowly fall out of favor as the growing data collection and search program at the LHC have continued to rule out the simplest BSM extensions of the SM. Instead, we take the more practical approach of asking what remaining corners of phase spaces accessible at the LHC could still be hiding potential new physics. These may include exotic final states yet unprobed but, as is pursued in this analysis, the possibility that new physics is buried in existing final states due to its mimicking of standard reconstructed objects.

3.3 The Extended Higgs Sector

For a BSM search to be viable, the BSM decay must be unique enough to have not been detected in existing measurements, yet be potentially large enough to be detectable in a dedicated, optimized search. As a relatively newly discovered particle, the Higgs sector remains an attractive search space for BSM physics, with current constraints on allowed Higgs to BSM couplings ranging between 20–60%, depending on assumptions [5]. In particular, extended, SM-neutral Higgs sectors, where BSM states couple only to the Higgs and not to the gauge fields directly, would not be ruled out by the more extensive searches performed for non-Higgs measurements that agree with the SM. Moreover, because of the small decay width of the Higgs boson, couplings to BSM states, even if small, can still lead to exotic decays with sizeable branching fractions that would be accessible at the LHC [6].

There are a few ways the SM Higgs sector could be extended. The simplest of these involves the addition of a (possibly complex) scalar or pseudoscalar (or both, if complex) singlet (SM+S), where scalar or pseudoscalar denotes whether the particle is even or odd under parity, respectively. Alternatively, one could introduce yet another Higgs field that is also a complex scalar and a doublet of the SU(2) group, known as two-Higgs-doublet models (2HDM). However, given the allowed degrees of freedom in these models, fairly restrictive constraints have already been imposed on their predictions [6]. Models with additional degrees of freedom, in which one introduces both a singlet and doublet (2HDM+S), are thus favored by current constraints. The above extensions, for instance, are well-motivated in theories of Minimal Composite Higgs Models (MCHM) and Next-to-Minimal Supersymmetric Standard Model (NMSSM).

Another important motivation for an extended Higgs sector is the production of an axion particle, as originally described by Pecci-Quinn theory [3], that serves as both a dark matter candidate and a resolution to the strong CP problem [4, 7, 8]. Such axion searches are also attractive astrophysically and cosmologically, due to the important role such a particle would play in early big bang nucleosynthesis, stellar formation, the cosmic microwave background radiation [9–11]. In most of these searches [1, 12, 13], masses $m_a \sim$ eV are typically sought, and production from the Higgs boson is generally not assumed or required. The most massive limits are typically of order $m_a \lesssim 10^2$ keV [14]. At the masses accessible at the LHC, however, $m_a \gtrsim 1$ MeV, the mechanisms needed to resolve the strong CP problem and other astrophysical and cosmological issues are not necessarily realized. Model assumptions thus need to be relaxed and the new state is typically referred to as an *axion-like* particle (ALP) instead [15]. In the interest of remaining model-agnostic, we will refer to this new, light (at least for LHC mass scales) state a. Regardless of model, an additional interaction vertex, of the form given in Fig. 3.8, is introduced.

Phenomenologically, whether the particle a is a scalar (even parity) or a pseudoscalar (odd parity) is relevant only insofar as calculating specific branching fractions [6]. Experimentally, it is of little consequence to the analysis whether one assumes one parity representation or the other. In ALP searches, since the axion interpretation of the particle a from Pecci-Quinn theory requires that it be a CP-odd pseudoscalar, the particle a is identified as a pseudoscalar. In the analysis we present, however, we remain agnostic toward the CP parity of a.

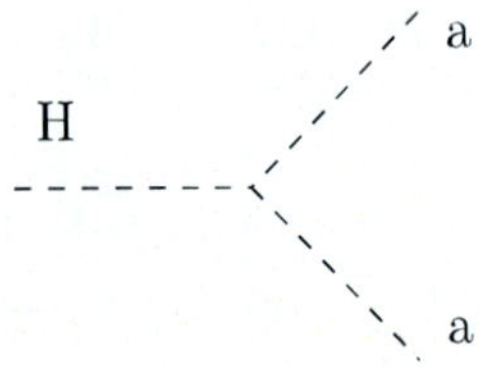

Fig. 3.8 The H → aa interaction vertex in the BSM extended Higgs sector

While the various models for H $\rightarrow$ aa production differ in their details and assumptions, there are a few common threads and phenomenological consequences for a decays. In general, the pseudoscalar inherits the same decay modes as the SM Higgs boson, meaning its coupling strengths are proportional to the mass of the decaying species, but with the potential for deviations. Thus, decays to quarks a $\rightarrow q\bar{q}$ are preferred above hadron production thresholds, while decays to lighter modes become more attractive below these thresholds. Under certain model assumptions, decays to lighter states like leptons a $\rightarrow$ ll and photons a $\rightarrow \gamma\gamma$ can become enhanced, or even dominant, even above hadron production thresholds [6]. Below $m_{\mathrm{a}} \lesssim 1\,\mathrm{GeV}$, theoretical estimates for a branching fraction tend to be plagued by large uncertainties as the non-perturbative QCD regime is crossed [6].

As the heavier decay modes are closed out, the decay width of the a particle is reduced, in turn, extending its lifetime. Thus, displaced and long-lived a decays are a general feature as m_{a} drops below hadron thresholds. The more a decay modes are closed out at low-m_{a}, the more this reduces the decay width of the a and extends its likely lifetime.

While H $\rightarrow$ aa$'$ production is also a distinct possibility, if, for instance, an additional pseudoscalar a$'$ (or other scalar) is allowed, we not consider this for the present study.

H → aa → 4γ A tantalizing feature of these low-m_{a} topologies is that their decay products are merged, obscuring their two-prong nature, and burying them in more mundane-looking SM signatures. In particular, for a $\rightarrow \gamma\gamma$, which is allowed at 1-loop (c.f. Fig. 3.7), in the low-m_{a} regime where this decay becomes attractive, the diphoton system becomes highly boosted and will often be misreconstructed as a single photon candidate. The H $\rightarrow$ aa $\rightarrow 4\gamma$ event would thus go largely undetected in existing SM H $\rightarrow \gamma\gamma$ measurements, making them challenging to pick out, especially if attempting to directly measure the m_{a} resonance.

While insufficient collision energy reach is a likely explanation for the lack of new physics signals at the LHC, an important, practical consideration that must be acknowledged is the possibility that new physics is manifest at the LHC but is simply buried in existing measurements due to current reconstruction algorithms being insensitive to them. While not much can be done about the former in the short-term, the latter presents a much more fruitful target in the short-term. It is for this reason the H $\rightarrow$ aa $\rightarrow 4\gamma$ decay presents an attractive topology for study, and is the focus of this thesis. While a number of other LHC experiment have probed direct a $\rightarrow \gamma\gamma$ production (i.e., a not produced from the Higgs boson), primarily from Pb-Pb collisions [16], this is the first attempt at CMS to look for H $\rightarrow$ aa $\rightarrow 4\gamma$ events and directly measure the mass spectrum of individual reconstructed photon candidates.

We focus on the a $\rightarrow \gamma\gamma$ decay mode for the mass range $m_{\mathrm{a}} \in 100\,\mathrm{MeV}$ to 1.2 GeV. While such decays are preferred to be long-lived, for this initial analysis, we assume that they decay promptly with negligible lifetime. A discussion of our results in light of relaxing this assumption is provided in Sect. 9.2.5. The relevant production thresholds at these masses are: the charmonium threshold at $m_{\mathrm{a}} \lesssim 3\,\mathrm{GeV}$, the tri-$\Pi^0$ threshold at $m_a \lesssim 405$ MeV, and the di-muon threshold at $m_a \lesssim 210$ MeV. If we allow charge symmetry to be violated, then the $a \rightarrow \Pi^0 + \gamma$ threshold at $m_a \lesssim 135$ MeV is important as well. Rough estimates for the $H \rightarrow aa \rightarrow 4\gamma$ cross section are discussed in Sect. 8.1.

3.3.1 H → aa → 4γ Phenomenology

This analysis focuses on the $H \rightarrow aa \rightarrow 4\gamma$ process with particle masses in the range $m_a = 0.1$ to 1.2 GeV so that both legs of the $H \rightarrow aa \rightarrow 4\gamma$ are reconstructed as single photon candidates, mimicking a SM $H \rightarrow \gamma\gamma$ event. An illustrative Feynman diagram for this process is shown in Fig. 3.9. We denote each single reconstructed photon corresponding to a merged $a \rightarrow \gamma\gamma$ as Γ_{reco}, or simply Γ.

The Higgs boson is produced resonantly at a mass of $m_H = 125$ GeV. In the case of the signal $H \rightarrow aa \rightarrow 4\gamma$, but also for the SM $H \rightarrow \gamma\gamma$, this means the invariant mass of the reconstructed photon legs, mGG, will peak sharply around the Higgs mass. This is in contrast to the dominant QCD backgrounds which are produced non-resonantly and thus exhibit a smoothly falling $m_{\Gamma\Gamma}$ spectrum. This can be exploited to significantly reduce non-resonant backgrounds. Since the Higgs is also produced primarily via gluon fusion, as opposed to QCD processes which are produced through one or more forms of quark annihilation, the angular distribution of its decay products also differ slightly from those of QCD. Exploiting this feature is major strategy of the diphoton event classifier used in the SM $H \rightarrow \gamma\gamma$ analysis, however, for simplicity in this first $H \rightarrow aa \rightarrow 4\gamma$ analysis, we do not take advantage of this effect.

As it is for the SM $H \rightarrow \gamma\gamma$ analysis, the primary background processes for the $H \rightarrow aa \rightarrow 4\gamma$ are those from QCD containing jets with neutral mesons decaying to photons. These are known as electromagnetically-enriched (EM) jets or photon "fakes". Because these are produced with cross-sections orders-of-magnitude larger than those of $H \rightarrow \gamma\gamma$ production (see Fig. 3.10), even though hadronization to EM-

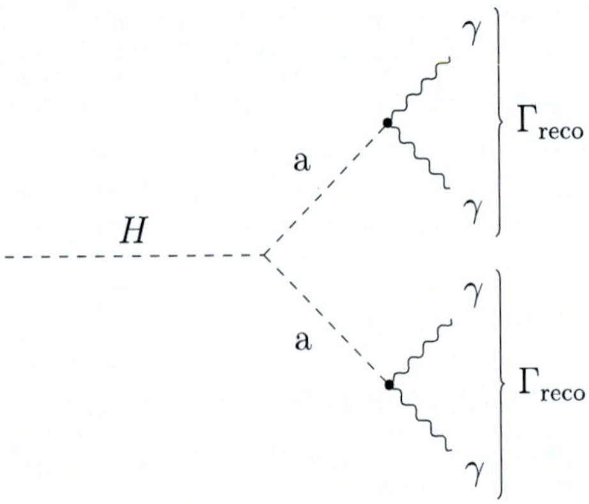

Fig. 3.9 Feynman diagram for the $H \rightarrow aa \rightarrow 4\gamma$ process. In the fully merged scenario, each $a \rightarrow \gamma\gamma$ decay is reconstructed as a single photon candidate Γ

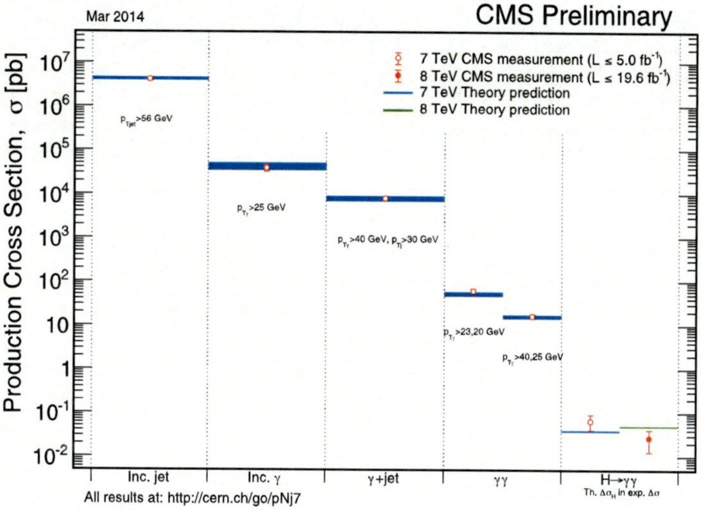

Fig. 3.10 Production cross sections for $\mathrm{H} \rightarrow \gamma\gamma$ and its leading background sources at the LHC for $\sqrt{s} = 8\,\mathrm{TeV}$. Credit: https://twiki.cern.ch/twiki/bin/view/CMSPublic/PhysicsResultsCombinedSMP

enriched jets is rare, they remain a major component of the reconstructed diphoton events, as seen in Fig. 3.11, even after signal selection criteria, described in Chap. 6, have been applied. These include QCD multijet and γ+jet production processes. Other background processes include prompt diphoton production and of course, the SM $\mathrm{H} \rightarrow \gamma\gamma$ itself.

The most notable yet subtle feature of $\mathrm{H} \rightarrow \mathrm{aa} \rightarrow 4\gamma$ events, and which presents the biggest challenge to detection, is the merged photon phenomenology of the $\mathrm{a} \rightarrow \gamma\gamma$ decay. Fundamentally, the collimation of the photons in the detector frame is determined by the angle between the a's boost vector and the decay axis of the diphotons, in the rest frame of the particle a. Nonetheless, to simplify the discussion, we parametrize the diphoton merging in terms of the Lorentz boost $\gamma_L = E_{\mathrm{a}}/m\mathrm{a}$ for a particle of energy E_{a} and mass $m\mathrm{a}$. The caveat, of course, is that different kinematic combinations of the diphoton 4-momenta can still lead to the same boost, such that the Lorentz boost does note uniquely determine the opening angle between the diphotons in the lab frame.

To appreciate the experimental challenge of performing such a measurement, it is helpful to visualize the typical generator-level opening angle between the two photons from the a in the detector frame. We use samples of $\mathrm{a} \rightarrow \gamma\gamma$ decays misreconstructed by PF as single photons passing selection criteria (see Sect. 6), taken from simulated $\mathrm{H} \rightarrow \mathrm{aa}$ events. The distribution of the generator-level opening angles between the leading (γ_1) and subleading (γ_2) photon in $p\mathrm{T}$ from the simulated a decay is shown in the left column of Fig. 3.12. The angles are expressed in number of ECAL crystals ($\Delta\phi \times \Delta\eta = 0.0174 \times 0.0174$) in the η direction, $\Delta\eta(\gamma_1, \gamma_2)^{\mathrm{gen}}$, versus the ϕ direction, $\Delta\varphi(\gamma_1, \gamma_2)^{\mathrm{gen}}$. Note that the collimation of the photons in the ECAL is

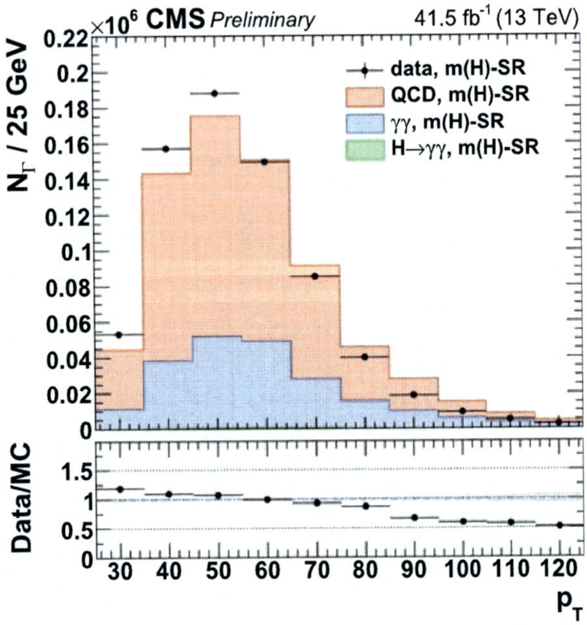

Fig. 3.11 Reconstructed pT distributions for 2017 data, with simulation used to show the contributions of events from QCD (multijet and γ + jet), $\gamma\gamma$, and the SM H $\rightarrow \gamma\gamma$. The event selection criteria are described in Chap. 6

ultimately determined by the opening angle between the diphoton axis and the a's boost vector in the rest frame of the a. The same γ_L can thus still lead to different apparent merging.

For all samples shown in Fig. 3.12, the a $\rightarrow \gamma\gamma$ is misreconstructed as a single photon candidate Γ. Two approximate merging thresholds are seen to occur at H $\rightarrow$ aa energies. The first occurs at roughly $m_{\text{a}} \sim 1.2$ GeV or $\gamma_L \sim 50$ and is roughly the boost threshold at which distinct photon showers begin to coalesce into a contiguous ECAL cluster. Distinct energy maxima can typically be resolved but because of the finite lateral granularity of the ECAL, (Molire radius ~ 1 ECAL crystal width) the diphoton showers begin to overlap. For reasons related to defining photon conversions as "single" photon candidates, such *shower merging*, within PF, still defines a single reconstructed photon. For decays which are barely resolved, conventional techniques can, in practice, be stretched to reconstruct m_{a}, although at a cost in reconstruction efficiency. In particular, at lower energies ($E \sim 1$ GeV), shower clustering tools become viable even for $\pi^0 \rightarrow \gamma\gamma$ mass reconstruction. Indeed, such tools are already in use for ECAL inter-crystal calibration [17]. ML-based techniques using shower shape and isolation variables as inputs, similar to those used for photon vs fake identification [18], could also be used. This threshold represents the upper range of m_{a} studied in this analysis. As boosts increase further in the shower-merged regime, a second merging threshold occurs at around $m_{\text{a}} \lesssim 200$ MeV or $\gamma_L \gtrsim 250$. At these boosts, the diphotons primarily deposit into the *same* ECAL crystal and are therefore no longer resolved in any sense of the word. While m_{a} regimes with such *instrumentally merged* signatures offer the most attractive branching fractions for a $\rightarrow \gamma\gamma$ detecion, experimentally, they are very difficult to probe as they require

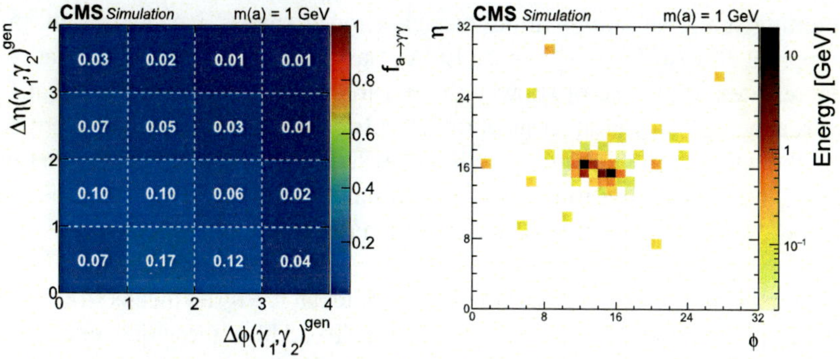

(a) Barely resolved, $m_a = 1\,\text{GeV}$, $\gamma_L = 50$

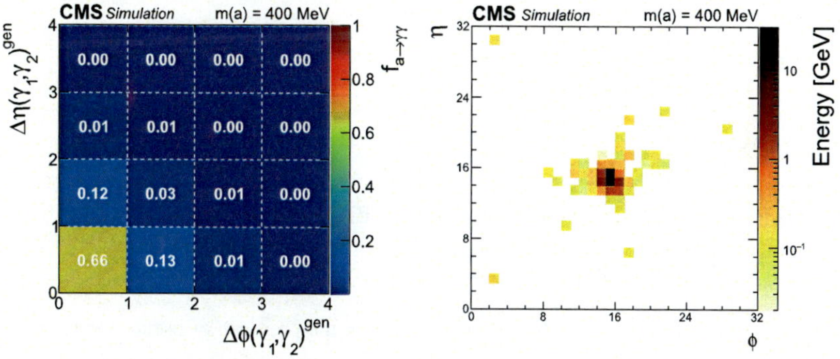

(b) Shower merged, $m_a = 400\,\text{MeV}$, $\gamma_L = 150$

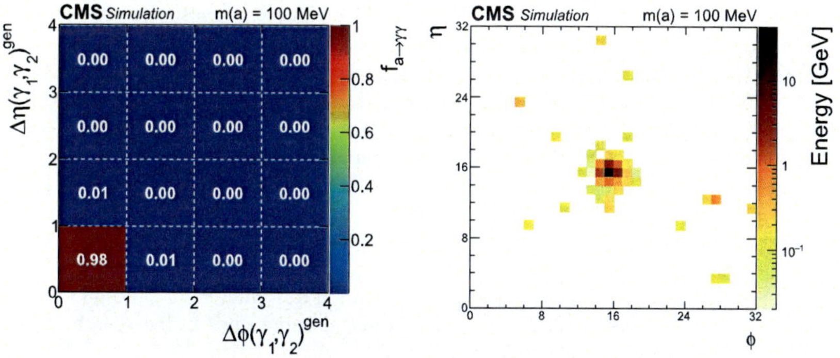

(c) Instrumentally merged, $m\;\; = 100\,\text{MeV}$, $\gamma_L = 625$

Fig. 3.12 Simulation results for the decay chain $\text{H} \to \text{aa}$, $\text{a} \to \gamma\gamma$ at various boosts: (upper plots) barely resolved, $m_a = 1\,\text{GeV}$, $\gamma_L = 50$; (middle plots) shower merged, $m_a = 400\,\text{MeV}$, $\gamma_L = 150$; and (lower plots) instrumentally merged, $m_a = 100\,\text{MeV}$, $\gamma_L = 625$. The left column shows the normalized distribution ($f_{a\to\gamma\gamma}$) of opening angles between the leading (γ_1) and subleading (γ_2) photons from the particle a decay, expressed by the number of crystals in the η direction, $\Delta\eta(\gamma_1, \gamma_2)_{\text{gen}}$, versus the ϕ direction, $\Delta\phi(\gamma_1, \gamma_2)_{\text{gen}}$. The right column displays the ECAL shower pattern for a single $\text{a} \to \gamma\gamma$ decay, plotted in relative ECAL crystal coordinates. In all cases, only decays reconstructed as a single PF photon candidate passing selection criteria are used exploiting subtle variations in the particle shower shape. For this reason, previous attempts at ATLAS [19–21], whose ECAL has a finer granularity than that of CMS, have resorted to the use of shower shape variables to parametrize the shape of the electromagnetic shower in terms of ratios and correlations between different detector cells about the energy maximum. While they have allowed some handle on this mass regime, they involve the art of trial-and-error and, historically, have only been able to select or "tag" signal-like objects rather than directly reconstruct the physical mass m_{a}.

In sum, while a direct measurement of m_{a} is ideal, it is by no means straightforward and an open question whether at all possible. To address these questions, we study what sensitivity is achievable with existing shower clustering and shower shape-based tools and show that we are led to consider all-new tools, and, indeed, to rethink the particle reconstruction workflow itself.

References

1. Zyla P et al (2020) Review of particle physics. PTEP 2020(8):083C01
2. Aaij R, Beteta CA, Ackernley T, Adeva B, Adinolfi M, Afsharnia H, Aidala CA, Aiola S, Ajaltouni Z, Akar S, Albrecht J, Alessio F, Alexander M, Albero AA et al (2021) Test of lepton universality in beauty-quark decays
3. Peccei RD, Quinn HR (1977) CP conservation in the presence of pseudoparticles. Phys Rev Lett 38:1440–1443
4. Peccei RD (2008) The strong CP problem and axions. In: Axions. Springer, Berlin, p 3
5. Sirunyan AM, Tumasyan A, Adam W, Ambrogi F, Asilar E, Bergauer T, Brandstetter J, Dragicevic M, Ero J, Del Valle AE et al (2019) Combined measurements of Higgs boson couplings in proton-proton collisions at $\sqrt{s} = 13$ TeV. Eur Phys J C 79:421
6. Curtin D, Essig R, Gori S, Jaiswal P, Katz A, Liu T, Liu Z, McKeen D, Shelton J, Strassler M, et al (2014) Exotic decays of the 125 gev higgs boson. Phys Rev D 90, Oct 2014
7. Essig R et al (2013) Dark sectors and new, light, weakly-coupled particles
8. Chadha-Day F, Ellis J, Marsh DJE (2021) Axion dark matter: What is it and why now?
9. Raffelt GG (2008) Astrophysical axion bounds. In: Axions, Springer, Berlin, p 51
10. Sikivie P (2008) Axion cosmology. In: Axions, Springer, Berlin, p 19
11. Marsh DJ (2016) Axion cosmology. Phys Rep 643:1
12. Graham PW, Irastorza IG, Lamoreaux SK, Lindner A, van Bibber KA (2015) Experimental searches for the axion and axion-like particles. Annu Rev Nucl Part Sci 65:485
13. Irastorza IG, Redondo J (2018) New experimental approaches in the search for axion-like particles. Prog Part Nucl Phys 102:89
14. Cadamuro D, Hannestad S, Raffelt G, Redondo J (2011) Cosmological bounds on sub-mev mass axions. J Cosmol Astropart Phys 2011:003
15. Bauer M, Neubert M, Thamm A (2017) Collider probes of axion-like particles. JHEP 12:044
16. Knapen S, Lin T, Lou HK, Melia T (2017) Searching for axionlike particles with ultraperipheral heavy-ion collisions. Phys Rev Lett 118:171801
17. Cavallari F et al (2006) CMS ECAL intercalibration of ECAL crystals using laboratory measurements
18. Khachatryan V et al (2014) Observation of the diphoton decay of the Higgs boson and measurement of its properties. Eur Phys J C 74(10):3076
19. Collaboration A (2012) Search for a Higgs boson decaying to four photons through light CP-odd scalar coupling using 4.9 fb^{-1} of 7 TeV *pp* collision data taken with ATLAS detector at the LHC 7

20. Aaboud M et al (2019) Search for pairs of highly collimated photon-jets in *pp* collisions at $\sqrt{s} = 13$TeV with the ATLAS detector. Phys Rev D 99:012008 Jan
21. Aad G, Abbott B, Abdallah J, Abdinov O, Aben R, Abolins M, AbouZeid OS, Abramowicz H, Abreu H et al (2016) Search for new phenomena in events with at least three photons collected in pp collisions at $\sqrt{s} = 8$ tev with the atlas detector. Eur Phys J C 76(4):1–26

Chapter 4
Analysis Strategy

In this Section, we introduce a novel technique that will enable the reconstruction of the mass of even the most highly-merged $a \rightarrow \gamma\gamma$ decays. This so-called end-to-end particle reconstruction technique is described in Sect. 4.1. In Sect. 4.2, we describe how this newfound mass reconstruction capability enables a direct discrimination of $H \rightarrow aa \rightarrow 4\gamma$ candidate events to be performed for the first time. Finally, in Sect. 4.3, we tie these techniques and strategies together into a coherent search for the $H \rightarrow aa \rightarrow 4\gamma$ decay.

4.1 End-to-End Particle Reconstruction

As described in the previous chapter, the merged $a \rightarrow \gamma\gamma$ topology presents a number of challenges that have hitherto prevented a direct measurement of the particle a mass at either CMS or ATLAS.

A natural choice for seeking to overcome these challenges is machine learning (ML)-based methods. A number of existing applications at CMS and ATLAS have used particle shower shapes or PF-based information information to tackle various discrimination [1–9] and reconstruction tasks [10]. While these tend to be the most common and straightforward uses of ML in high energy physics, as we show in Sect. 7, even these are inadequate to tackle merged $a \rightarrow \gamma\gamma$ mass reconstruction. Evidently, the use of machine-learning methods, on their own, does not guarantee improved reconstruction sensitivity. It is important, therefore, to acknowledge that ML is simply a tool for extracting information, not creating it.

What must be given equal consideration—perhaps even more so—is the choice of inputs given to the ML algorithms. For many, mostly historical reasons, PF and shower shapes tend to be the starting point of most ML applications at CMS or ATLAS. However, with the emergence of modern ML, or "deep learning", the opportunity arises for a much more fundamental change in the particle reconstruction

M. Andrews, *Search for Exotic Higgs Boson Decays to Merged Diphotons*,
Springer Theses, https://doi.org/10.1007/978-3-031-25091-0_4

strategy itself, a change that, we show, provides a breakthrough for a $\to \gamma\gamma$ mass reconstruction.

As has been learned in industry [11–14], and in simpler physics experiments [15–17], it is critical to use as raw and rich a form of inputs as possible, in order to realize the breakthrough feature extraction capabilities of modern ML algorithms. This motivates the choice of a so-called "end-to-end" particle reconstruction strategy wherein one bypasses all unnecessary, intermediate steps as much as possible, and allows the ML algorithm to train directly on minimally-processed detector data with the objective of regressing the final, desired quantity of interest. This is achieved by casting the detector data as high-fidelity images and using these to train a convolutional neural network (CNN) [18–20] that outputs an estimate of the final parent particle property of interest, in this case, the particle mass, m_{a}. Order-invariant, graph-based ML models can be also be used [21].

This has the following potential advantages. First, is the gain in information granularity offered by detector data for features that cannot be easily reduced to a particle-level, or even shower shape type representations. Second, the choice of a CNN, or any similar hierarchical ML architecture such as a graph-based network, allows to learn detector features across several length scales, from the crystal-level to the cluster-level and beyond, in a synergistic way. This is particularly advantageous for boosted decays that may exhibit merging at multiple scales. Finally, by training on minimally-processed data rather than heavily filtered or clustered data, the ML algorithm learns to adapt to more varied, higher-dimensional changes in the data, developing a greater robustness to evolving data-taking conditions.

In this analysis, therefore, we employ a novel, end-to-end particle reconstruction-based strategy for directly reconstructing the particle mass, m_{a}. In Sect. 7, we show that this technique achieves substantial sensitivity gain allowing to directly probe previously inaccessible boost regimes of the a $\to \gamma\gamma$ decay. A first-of-its kind direct measurement of the merged photon mass in this analysis is thus enabled by this novel technique. The technique builds on earlier work in developing the end-to-end ML technique for discriminating electrons vs photons [22], H $\to \gamma\gamma$ vs backgrounds [19], and quark versus gluon-jets [18].

To simplify the scope of the end-to-end ML technique in this first analysis application, only ECAL barrel photons are considered in this analysis. The impact of this on signal selection efficiency is discussed in Chap. 6.

4.2 H $\to$ aa $\to 4\gamma$ Discrimination

Using the end-to-end particle reconstruction technique, we discriminate H $\to$ aa $\to 4\gamma$ candidates by directly reconstructing the merged photon mass spectrum m_Γ, for each assumed a $\to \gamma\gamma$ candidate Γ in each event, within the region-of-interest $m_\Gamma \in (0., 1.2)$ GeV. This represents the first attempt at the LHC to directly measure the merged a $\to \gamma\gamma$ mass spectrum.

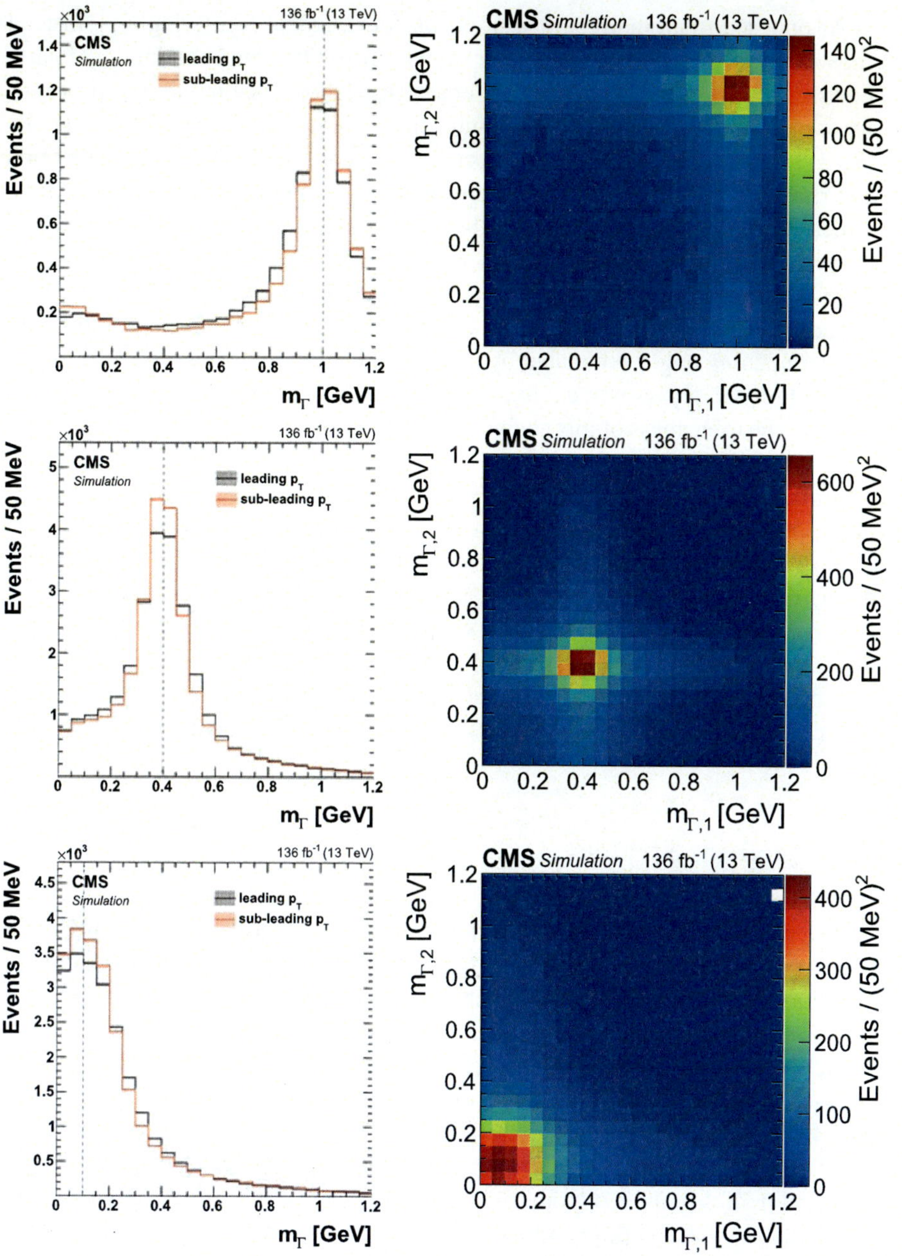

Fig. 4.1 Regressed 1D (left column) and 2D (right column) m_Γ distributions for simulated H → aa → 4γ events passing selection requirements at m_a = 100 MeV (bottom row), 400 MeV (middle row), and 1 GeV (top row)

As opposed to an indirect measurement–for instance, measuring $m_{\Gamma\Gamma}$ instead—this will allow for a potentially higher signal sensitivity over a continuum of particle a masses. In the event of a detection, observing a resonance in the m_{Γ} spectrum, as opposed to an excess in an artificial signal discriminant, e.g., a multivariate classifier-based "signal score", provides a more compelling, physically intuitive result. For this first analysis, the measurement is performed under the assumption that the particle a decays promptly. However, given the limited lifetime resolution of the ECAL to merged photons, we do expect the m_{Γ} regression to still be sensitive to the displaced decay vertices of long-lived as, for detector-frame decay lengths of $cT \lesssim 40$cm, or decay lengths of $c\tau \lesssim 0.15$cm in the $\mathrm{a} \to \gamma\gamma$ rest frame. However, due to the smaller opening angle subtended by a displaced decay vertex at the ECAL surface, such long-lived decays will exhibit a mass spectrum gradually skewed toward lower masses. The robustness of this analysis to the long-lived scenario is discussed further at the end of Sect. 9.1.

Furthermore, we perform a two-dimensional measurement of the merged photon mass spectrum $m_{\Gamma,2}$ (subleading p_{T}) vs. $m_{\Gamma,1}$ (leading p_{T}), 2D-m_{Γ} , one for each $\mathrm{a} \to \gamma\gamma$ leg of the candidates $\mathrm{H} \to \mathrm{aa}$ candidate event. Under the assumption that the particle a pairs in the hypothetical $\mathrm{H} \to \mathrm{aa} \to 4\gamma$ event have identical mass, this will result in the 2-D mass peak of signal events lying along the "diagonal" of the 2D-m_{Γ} distribution, as illustrated in Fig. 4.1. This feature can be exploited to significantly reduce the dominant background contributions from QCD which, in general, will not have correlated masses between photons and/or jets.

4.3 Analysis Strategy

The analysis starts by defining as loose a photon candidate selection as possible to maximize the $\mathrm{a} \to \gamma\gamma$ selection efficiency. A diphoton event topology is then required. Each reconstructed photon candidate Γ is then used to fill a 2D-m_{Γ} distribution or "template" to serve either as a signal model or a background model. A separate signal model is generated for each m_{a} hypotheses in the range $m_{\mathrm{a}} = [0.1, 1.2]$ GeV. Representative distributions are shown in the right column of Fig. 4.1.

The signal model is derived from simulated $\mathrm{H} \to \mathrm{aa} \to 4\gamma$ events. Under the identical particle a mass hypothesis, the nominal signal inhabits the diagonal region of the 2D-m_{Γ} plane. Additionally, we assume the a strictly arises from Higgs boson decays with no associated invisible decay, allowing us to define an orthogonal signal region within the Higgs boson mass window. The signal (SR) and sideband (SB) regions along the $m_{\Gamma\Gamma}$ axis are thus,

- m_H-SB$_{\mathrm{low}}$: $100 < m_{\Gamma\Gamma} < 110$ GeV
- m_H-SR : $110 < m_{\Gamma\Gamma} < 140$ GeV
- m_HSB$_{\mathrm{high}}$: $140 < m_{\Gamma\Gamma} < 180$ GeV.

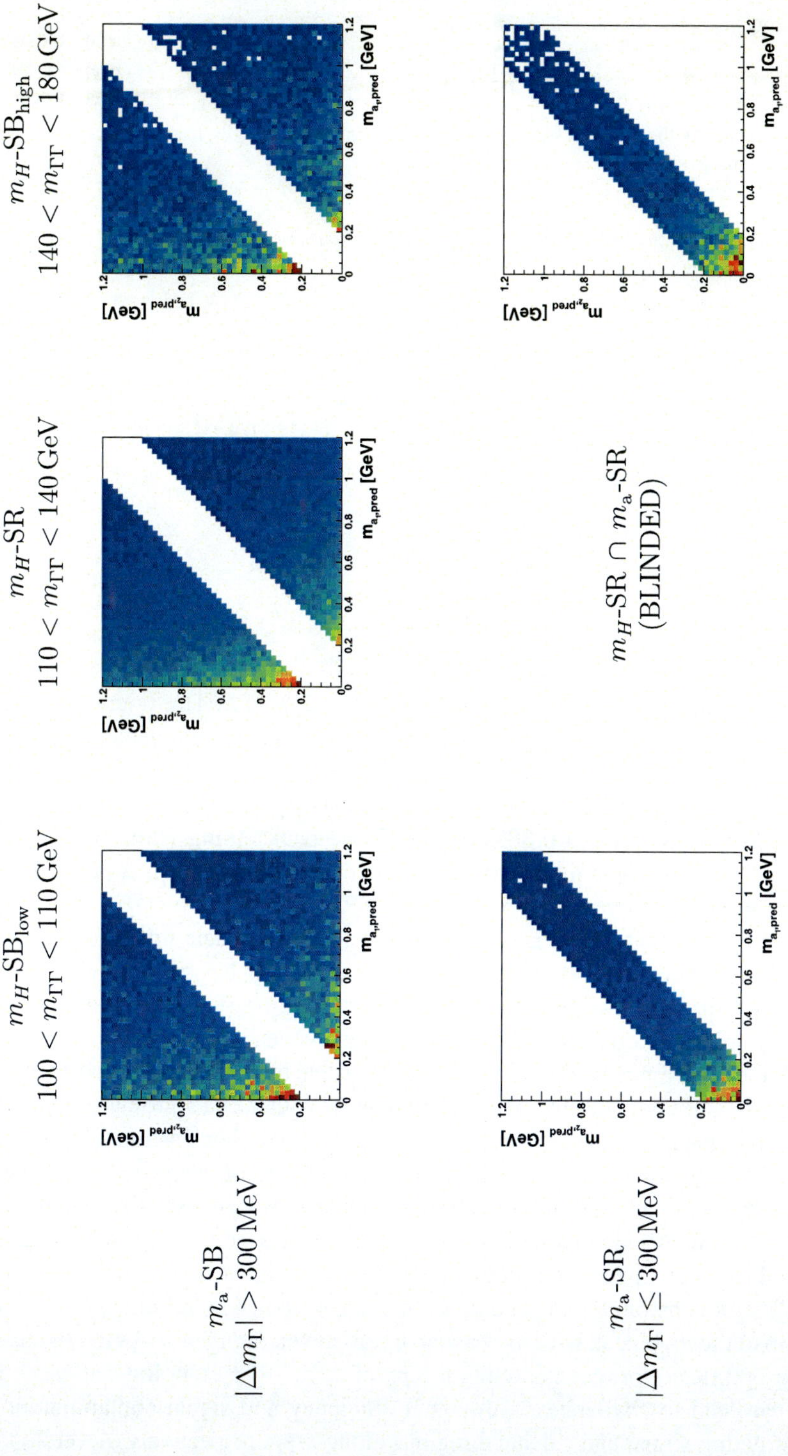

Fig. 4.2 Illustration of the different data regions in the $(m_{\Gamma\Gamma}, m_{\text{a}})$ plane and the representative shape of their corresponding 2D-m_{Γ} templates

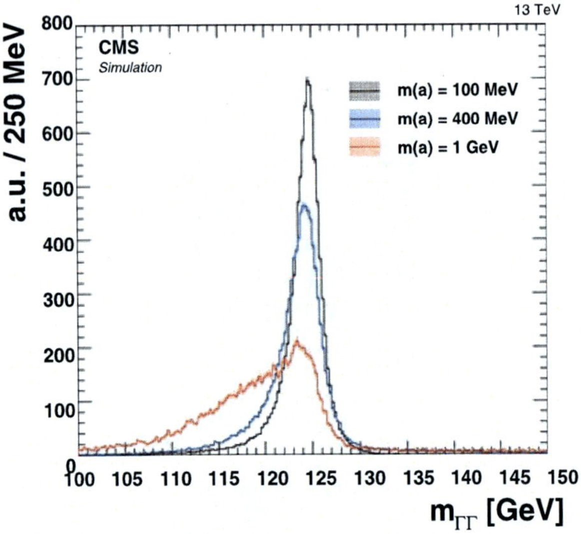

Fig. 4.3 Degradation of the reconstructed $m_{\Gamma\Gamma}$ with increasing m_a for simulated $H \to aa \to 4\gamma$ events passing selection requirements, at various generated m_a. As the $a \to \gamma\gamma$ opening angle increases with m_a, the PF-reconstructed p_T becomes increasingly underestimated, in turn, impacting the reconstructed $m_{\Gamma\Gamma}$

While the regions along the 2D-m_Γ axis are

- m_a-SB : $\Delta m_a > 300$ MeV ("2D-m_Γ off-diagonal")
- m_a-SR : $\Delta m_a \leq 300$ MeV ("2D-m_Γ diagonal").

The different data regions are illustrated in Fig. 4.2. The boundaries between regions are optimized for minimal signal contamination in the sideband regions while still maintaining sufficient statistics. In $m_{\Gamma\Gamma}$, this results in between 87% ($m_a \approx 1$ GeV) to 99% ($m_a \approx 100$ MeV) of signal events falling within the m_H-SR. In 2D-m_Γ, this results in the top 80% of the 2D-m_Γ peak falling within the m_a-SR, for all considered m_a. The final signal region over which the search is performed is the intersection of the two signal regions, m_H-SR $\cap$ m_a-SR.

The background model is derived primarily through data-driven techniques that exploit the data sidebands of the above signal regions to obtain an estimate of the total background contribution in the final signal region. The contribution of the SM $H \to \gamma\gamma$, however, is derived using simulated events.

To measure the presence of a signal, a statistical hypothesis test is performed [23], comparing the compatibility of the observed data with the background-only versus the signal plus background hypotheses. The maximum likelihood estimation method [24] is used to construct a test statistic for comparing the compatibility of the data with the different hypotheses and for extracting the best-fit signal strength. In the event of a detection, the significance of the signal is calculated, otherwise the CL_s metric [24] is used to set an upper limit on excluded signal strengths.

A major constraint to expanding the m_a range studied in this analysis is the degradation of the $m_{\Gamma\Gamma}$ peak with increasing m_a, as shown in Fig. 4.3. As the PF algorithm is not optimized for reconstructing the p_T of the $a \to \gamma\gamma$ topology, at $m_a \gtrsim 1$ GeV, this can lead to challenges with signal efficiency and signal contamination when constructing signal and sideband regions along $m_{\Gamma\Gamma}$, respectively. A version of this analysis utilizing end-to-end $p_{T,a}$ reconstruction to overcome this effect at higher m_a is being considered for future work. An exploratory study describing the potential of this application is given at the end of our conclusions, in Chap. 10.

References

1. Sirunyan AM et al (2017) Identification of heavy-flavour jets with the CMS detector in pp collisions at 13 TeV. Identification of heavy-flavour jets with the CMS detector in pp collisions at 13 TeV. JINST 13:P05011 (114 p, Dec 2017)
2. Collaboration CMS (2019) A deep neural network to search for new long-lived particles decaying to jets
3. Kasieczka G, Plehn T, Russell M, Schell T (2017) Deep-learning top taggers or the end of QCD? JHEP 05:006
4. Aad G (2018) Performance of top-quark and W-boson tagging with ATLAS in Run 2 of the LHC Eur Phys J C 79:375 (79 p, Aug 2018). 79 pages in total, author list starting page 63, 39 figures, 6 tables, submitted to The European Physical Journal C. All figures including auxiliary figures are available at http://atlas.web.cern.ch/Atlas/GROUPS/PHYSICS/PAPERS/JETM-2018-03
5. Qu H, Gouskos L (2020) Jet tagging via particle clouds. Phys Rev D 101:056019 Mar
6. Butter A, Kasieczka G, Plehn T, Russell M (2018) Deep-learned top tagging with a Lorentz layerSciPost Phys 5
7. Louppe G, Cho K, Becot C, Cranmer K (2019) QCD-aware recursive neural networks for jet physics. JHEP 01:057
8. Ju X, Nachman B (2020) Supervised jet clustering with graph neural networks for Lorentz boosted bosons. Phys Rev D 102:075014 Oct
9. Komiske PT, Metodiev EM, Thaler J (2019) Energy flow networks: deep sets for particle jets. J High Energy Phys 2019(1):121
10. Sirunyan AM et al (2020) A deep neural network for simultaneous estimation of b quark energy and resolution. Comput Softw Big Sci 4:10
11. Krizhevsky A, Sutskever I, Hinton GE (2017) Imagenet classification with deep convolutional neural networks. Commun ACM 60:84–90 May
12. Esteva A, Kuprel B, Novoa R et al (2017) Dermatologist-level classification of skin cancer with deep neural networks. Nature 542:115–118
13. Silver D, Hubert T, Schrittwieser J, Antonoglou I, Lai M, Guez A, Lanctot M, Sifre L, Kumaran D, Graepel T, Lillicrap T, Simonyan K, Hassabis D (2017) Mastering chess and shogi by self-play with a general reinforcement learning algorithm
14. Senior A, Evans R, Jumper J et al (2020) Improved protein structure prediction using potentials from deep learning. Nature 577:706–710
15. Aurisano A, Radovic A, Rocco D, Himmel A, Messier MD, Niner E, Pawloski G, Psihas F, Sousa A, Vahle P (2016) A convolutional neural network neutrino event classifier. JINST 11(09):P09001
16. Adams C et al (2019) Deep neural network for pixel-level electromagnetic particle identification in the microboone liquid argon time projection chamber. Phys Rev D 99:092001 May
17. Uboldi L, Ruth D, Andrews M, Wang MHLS, Wenzel H, Wu W, Yang T (2021) Extracting low energy signals from raw LArTPC waveforms using deep learning techniques—A proof of concept
18. Andrews M, Alison J, An S, Burkle B, Gleyzer S, Narain M, Paulini M, Poczos B, Usai E (2020) End-to-end jet classification of quarks and gluons with the CMS open data. In: Nuclear instruments and methods in physics research section A: accelerators, spectrometers, detectors and associated equipment, p 164304

19. Andrews M, Paulini M, Gleyzer S, Poczos B (2020) End-to-end physics event classification with CMS open data: applying image-based deep learning to detector data for the direct classification of collision events at the LHC. Comput Softw Big Sci 4(1):6
20. De Oliveira L, Nachman B, Paganini M (2020) Electromagnetic showers beyond shower shapes. Nucl Instrum Meth A 951:162879
21. Ju X, Farrell S, Calafiura P, Murnane D, Prabhat, Gray L, Klijnsma T, Pedro K, Cerati G, Kowalkowski J, Perdue G, Spentzouris P, Tran N, Vlimant J-R, Zlokapa A, Pata J, Spiropulu M, An S, Aurisano A, Hewes J, Tsaris A, Terao K, Usher T (2020) Graph neural networks for particle reconstruction in high energy physics detectors
22. Andrews M, Paulini M, Gleyzer S, Poczos B (2019) Exploring end-to-end deep learning applications for event classification at CMS. In: EPJ web of conferences, 01 2019, vol 214, p 06031
23. Cowan G, Cranmer K, Gross E, Vitells O (2011) Asymptotic formulae for likelihood-based tests of new physics. Eur Phys J C 71
24. The ATLAS Collaboration, The CMS Collaboration (2011) Procedure for the LHC Higgs boson search combination in Summer 2011. Technical report CMS-NOTE-2011-005. ATL-PHYS-PUB-2011-11, CERN, Geneva, Aug 2011

Chapter 5
Data Sets

The data samples used in this analysis, together with their simulated equivalents, correspond to pp collision events collected by the CMS detector during the Run II phase of the LHC from 2016–2018 at a center-of-mass collision energy of $\sqrt{s} = 13$ TeV. This chapter documents the data set names used for both recorded data, given in Sect. 5.1, and simulated data, given in Sect. 5.2. For the simulated data, an overview of their use in this analysis, as well as the particulars of their generation, are provided.

5.1 Recorded Data

The recorded data sets are composed of events reconstructed from the 2016-2018 era CMS Re-Reco campaigns. They represent a combined integrated luminosity of $\int \mathcal{L}\,\mathrm{dt} = 137\,\mathrm{fb}^{-1}$ of data certified for physics analysis. Due to a number of AOD data tier files being inaccessible, particularly in the 2018 era data sets, the actual integrated luminosity used in this analysis corresponds to $\int \mathcal{L}\,\mathrm{dt} = 136\,\mathrm{fb}^{-1}$. The complete list of data set names are given in Table 5.1 with corresponding good run list names and luminosities in Table 5.2.

Both MINIAOD and parent AOD data sets are used: in situations where the a $\rightarrow \gamma\gamma$ decays into distinct clusters in the ECAL ($m_{\mathrm{a}} \gtrsim 1$ GeV), the softer of the two clusters may be dropped from the MINIAOD data set due to e/γ pruning [1] and regressed as a photon. Therefore, to maximize signal efficiency, while the event selection is performed using MINIAOD-level quantities—as is standard practice—the ECAL detector inputs used for the actual m_{a} regression are sourced from the parent AOD data set.

M. Andrews, *Search for Exotic Higgs Boson Decays to Merged Diphotons*, Springer Theses, https://doi.org/10.1007/978-3-031-25091-0_5

Table 5.1 Recorded data sample names by era for the years 2016–2018

Data set name
/DoubleEG/Run2016B-17Jul2018_ver2-v1/MINIAOD
/DoubleEG/Run2016B-07Aug17_ver2-v2/AOD
/DoubleEG/Run2016C-17Jul2018-v1/MINIAOD
/DoubleEG/Run2016C-07Aug17-v1/AOD
/DoubleEG/Run2016D-17Jul2018-v1/MINIAOD
/DoubleEG/Run2016D-07Aug17-v1/AOD
/DoubleEG/Run2016E-17Jul2018-v1/MINIAOD
/DoubleEG/Run2016E-07Aug17-v1/AOD
/DoubleEG/Run2016F-17Jul2018-v1/MINIAOD
/DoubleEG/Run2016F-07Aug17-v1/AOD
/DoubleEG/Run2016G-17Jul2018-v1/MINIAOD
/DoubleEG/Run2016G-07Aug17-v1/AOD
/DoubleEG/Run2016H-17Jul2018-v1/MINIAOD
/DoubleEG/Run2016H-07Aug17-v1/AOD
/DoubleEG/Run2017B-31Mar2018-v1/MINIAOD
/DoubleEG/Run2017B-17Mar2017-v1/AOD
/DoubleEG/Run2017C-31Mar2018-v1/MINIAOD
/DoubleEG/Run2017C-17Mar2017-v1/AOD
/DoubleEG/Run2017D-31Mar2018-v1/MINIAOD
/DoubleEG/Run2017D-17Mar2017-v1/AOD
/DoubleEG/Run2017E-31Mar2018-v1/MINIAOD
/DoubleEG/Run2017E-17Mar2017-v1/AOD
/DoubleEG/Run2017F-31Mar2018-v1/MINIAOD
/DoubleEG/Run2017F-17Mar2017-v1/AOD
/EGamma/Run2018A-17Sep2018-v2/MINIAOD
/EGamma/Run2018A-17Sep2018-v2/AOD
/EGamma/Run2018B-17Sep2018-v1/MINIAOD
/EGamma/Run2018B-17Sep2018-v1/AOD
/EGamma/Run2018C-17Sep2018-v1/MINIAOD
/EGamma/Run2018C-17Sep2018-v1/AOD
/EGamma/Run2018D-17Sep2018-v2/MINIAOD
/EGamma/Run2018D-22Jan2019-v2/AOD

Table 5.2 Lists of certified physics data and their corresponding integrated luminosity

Data set name	$\int \mathcal{L}\,\mathrm{dt}$ [fb^{-1}]
Cert_271036-284044_13TeV_ReReco_07 Aug2017_Collisions16_JSON	36.32 (36.25)
Cert_294927-306462_13TeV_EOY2017 ReReco_Collisions17_JSON	41.53
Cert_314472-325175_13TeV_17 SeptEarlyReReco2018ABC_PromptEraD_Collisions18_JSON	59.74 (58.75)
Total RunII	137.6 (136.5)

Values in parenthesis indicate actual luminosities obtained after accounting for inaccessible AOD files (missing lumi $\approx 0.8\%$). Luminosity uncertainties are in the third significant figure (see Sect. 8.3)

5.2 Simulated Data

Simulated samples generated using Monte-Carlo methods (MC) are primarily used to derive the signal model but are also relevant for studying the different signal and background processes and optimizing the analysis strategy accordingly. Separate simulated samples are produced for each year of data taking with simulation parameters tuned to each year's particular conditions where possible.

5.2.1 *Signal Samples*

Simulated $\mathrm{H} \to \mathrm{aa} \to 4\gamma$ samples generated with $m_{\mathrm{a}} = 0.1, 0.2, 0.4, 0.6, 0.8, 1.0, 1.2$ GeV are directly used to fill the templates that define the signal model (see Sect. 8.1). An interpolation procedure, described in Sect. 8.1, is used to generate additional samples at 0.1 GeV intervals of m_{a}. For all events, the hard-scatter process $\mathrm{H} \to \mathrm{aa} \to 4\gamma$ is generated with `MADGRAPH5_aMC@NLO` at leading-order (LO) using the `SM + Dark Vector + Dark Higgs` model [2, 3] with upto one associated parton and negligible a lifetime. The basic phenomenology of the decay, however, is model-independent for a given a mass and lifetime. The Higgs boson is produced inclusively of all SM production modes. The generated hard process is then interfaced with `Pythia8` for parton matching and hadronization. The Parton distribution functions (PDFs) are taken from the NNPDF3.1 set. Underlying event for the *pp* collision uses the `CP5` (`CUETP8M1`) tune for 2017–2018 (2016) samples. The output is then interfaced with `GEANT4` to simulate the detailed detector response of CMS using the geometry and conditions relevant for each year.

The effects of multiple *pp* interactions other than from that containing the hard-scattering event, or pileup interactions, are also simulated. This is done by overlaying minimum bias events onto the existing output from the primary *pp* interaction. Pileup interactions from both the nominal bunch crossing (in-time pileup), or from earlier or later bunch crossings (out-of-time pileup) are accounted for. The distribution in

Table 5.3 Simulated $H \rightarrow aa \rightarrow 4\gamma$ signal samples

Data set name	m(a) (GeV)
/HAHMHToAA_AToGG_MA-0p1GeV_TuneCUETP8M1_PSweights_13TeV-madgraph_pythia8/	0.1
/HAHMHToAA_AToGG_MA-0p2GeV_TuneCUETP8M1_PSweights_13TeV-madgraph_pythia8/	0.2 GeV
/HAHMHToAA_AToGG_MA-0p4GeV_TuneCUETP8M1_PSweights_13TeV-madgraph_pythia8/	0.4
/HAHMHToAA_AToGG_MA-0p6GeV_TuneCUETP8M1_PSweights_13TeV-madgraph_pythia8/	0.6
/HAHMHToAA_AToGG_MA-0p8GeV_TuneCUETP8M1_PSweights_13TeV-madgraph_pythia8/	0.8
/HAHMHToAA_AToGG_MA-1GeV_TuneCUETP8M1_PSweights_13TeV-madgraph_pythia8/	1
/HAHMHToAA_AToGG_MA-1p2GeV_TuneCUETP8M1_PSweights_13TeV-madgraph_pythia8/ RunIISummer16MiniAODv3-PUMoriond17_94X_mcRun2_asymptotic_v3-v1/MINIAODSIM	1.2
RunIISummer16DR80Premix-PUMoriond17_80X_mcRun2_asymptotic_2016_TrancheIV_v6-v1/AODSIM	
/HAHMHToAA_AToGG_MA-0p1GeV_TuneCP5_PSweights_13TeV-madgraph_pythia8/	0.1
/HAHMHToAA_AToGG_MA-0p2GeV_TuneCP5_PSweights_13TeV-madgraph_pythia8/	0.2
/HAHMHToAA_AToGG_MA-0p4GeV_TuneCP5_PSweights_13TeV-madgraph_pythia8/	0.4
/HAHMHToAA_AToGG_MA-0p6GeV_TuneCP5_PSweights_13TeV-madgraph_pythia8/	0.6
/HAHMHToAA_AToGG_MA-0p8GeV_TuneCP5_PSweights_13TeV-madgraph_pythia8/	0.8
/HAHMHToAA_AToGG_MA-1GeV_TuneCP5_PSweights_13TeV-madgraph_pythia8/	1
/HAHMHToAA_AToGG_MA-1p2GeV_TuneCP5_PSweights_13TeV-madgraph_pythia8/	1.2
RunIIFall17MiniAODv2-PU2017_12Apr2018_94X_mc2017_realistic_v14-v1/MINIAODSIM	
RunIIFall17DRPremix-PU2017_94X_mc2017_realistic_v11-v1/AODSIM	
/HAHMHToAA_AToGG_MA-0p1GeV_TuneCP5_PSweights_13TeV-madgraph_pythia8/	0.1
/HAHMHToAA_AToGG_MA-0p2GeV_TuneCP5_PSweights_13TeV-madgraph_pythia8/	0.2
/HAHMHToAA_AToGG_MA-0p4GeV_TuneCP5_PSweights_13TeV-madgraph_pythia8/	0.4
/HAHMHToAA_AToGG_MA-0p6GeV_TuneCP5_PSweights_13TeV-madgraph_pythia8/	0.6
/HAHMHToAA_AToGG_MA-0p8GeV_TuneCP5_PSweights_13TeV-madgraph_pythia8/	0.8
/HAHMHToAA_AToGG_MA-1GeV_TuneCP5_PSweights_13TeV-madgraph_pythia8/	1
/HAHMHToAA_AToGG_MA-1p2GeV_TuneCP5_PSweights_13TeV-madgraph_pythia8/	1.2
RunIIAutumn18MiniAOD-102X_upgrade2018_realistic_v15-v2/MINIAODSIM	
RunIIAutumn18DRPremix-102X_upgrade2018_realistic_v15-v2/AODSIM	

Table 5.4 Simulated background samples, Run 2016

Data set name
/GluGluHToGG_M125_13TeV_amcatnloFXFX_pythia8/
RunIISummer16MiniAODv3-PUMoriond17_94X_mcRun2_asymptotic_v3_ext2-v2/MINIAODSIM
RunIISummer16DR80Premix-PUMoriond17_80X_mcRun2_asymptotic_2016_TrancheIV_v6_ext2-v1/AODSIM
/DYToEE_NNPDF30_13TeV-powheg-pythia8/
RunIISummer16MiniAODv2-EGM0_80X_mcRun2_asymptotic_2016_TrancheIV_v6-v1/MINIAODSIM
RunIISummer16DR80Premix-EGM0_80X_mcRun2_asymptotic_end2016_forEGM_v0-v1/AODSIM

Table 5.5 Simulated background samples, Run 2017

Data set name
/GluGluHToGG_M125_13TeV_amcatnloFXFX_pythia8/
RunIIFall17MiniAODv2-PU2017_12Apr2018_94X_mc2017_realistic_v14-v1/MINIAODSIM
RunIIFall17DRPremix-94X_mc2017_realistic_v10-v1/AODSIM
/DiPhotonJets_MGG-80toInf_13TeV_amcatnloFXFX_pythia8/
RunIIFall17MiniAODv2-PU2017_12Apr2018_94X_mc2017_realistic_v14-v1/MINIAODSIM
/GJet_Pt-20to40_DoubleEMEnriched_MGG-80toInf_TuneCP5_13TeV_Pythia8/
RunIIFall17MiniAODv2-PU2017_12Apr2018_94X_mc2017_realistic_v14-v1/MINIAODSIM
/GJet_Pt-40toInf_DoubleEMEnriched_MGG-80toInf_TuneCP5_13TeV_Pythia8/
RunIIFall17MiniAODv2-PU2017_12Apr2018_94X_mc2017_realistic_v14-v2/MINIAODSIM
/QCD_Pt-30to40_DoubleEMEnriched_MGG-80toInf_TuneCP5_13TeV_Pythia8/
RunIIFall17MiniAODv2-PU2017_12Apr2018_94X_mc2017_realistic_v14-v1/MINIAODSIM
/QCD_Pt-40toInf_DoubleEMEnriched_MGG-80toInf_TuneCP5_13TeV_Pythia8/
RunIIFall17MiniAODv2-PU2017_12Apr2018_94X_mc2017_realistic_v14-v1/MINIAODSIM
RunIIFall17MiniAODv2-PU2017_12Apr2018_94X_mc2017_realistic_v14_ext1-v1/MINIAODSIM
RunIIFall17DRPremix-94X_mc2017_realistic_v10_ext1-v1/AODSIM
/DYToEE_M-50_NNPDF31_13TeV-powheg-pythia8/
RunIIFall17MiniAODv2-PU2017_12Apr2018_94X_mc2017_realistic_v14-v1/MINIAODSIM
RunIIFall17DRPremix-94X_mc2017_realistic_v10-v1/AODSIM

number of pileup interactions used in simulation corresponds to that projected in data for the relevant year. A pileup re-weighting procedure (see Sect. 8.1) is then applied to the simulated events to correct for any residual differences in the observed pileup distribution. The average number of pileup interactions measured in data ranges between 23 and 37 for 2016 and 2018.

Note that the simulated $\mathrm{H} \rightarrow \mathrm{aa} \rightarrow 4\gamma$ samples are not used to train the m_Γ regressor, which instead uses simulated single $\mathrm{a} \rightarrow \gamma\gamma$ decays, with continuously distributed m_a [1]. More information about the samples used to train the m_Γ regressor is given in Sect. 7.2.

The full list of simulated signal samples used for the main $\mathrm{H} \rightarrow \mathrm{aa} \rightarrow 4\gamma$ signal search is provided in Table 5.3.

5.2.2 *Background Samples*

The simulated background samples are used to optimize various aspects of the event selection and background estimation procedure and to better understand the response of the m_a regressor to photon and jet backgrounds. With the exception of the $\mathrm{H} \rightarrow \gamma\gamma$ sample, simulated samples are not directly used to estimate background contributions for this analysis, which are instead derived from data, as described in Sect. 8.2.

The relevant background samples are similar to those used for the dedicated $\mathrm{H} \rightarrow \gamma\gamma$ analysis. In order of decreasing cross-section, these are: QCD multijet and γ + jet production, prompt diphoton production, and $\mathrm{H} \rightarrow \gamma\gamma$ production.

The QCD samples are generated with `Pythia8` at LO with phase space cuts that maximize overlap with that of (resonant) Higgs boson production. These are then further passed through physics object filters to preferentially select events containing electromagnetic-enriched (EM) jets, which have photon-like signatures–typically isolated jets containing one or more merged $\pi^0/\eta \rightarrow \gamma\gamma$ (see discussion in Appendix A.1). These are also intended to maximize the phase space overlap with $\mathrm{H} \rightarrow \gamma\gamma$. Despite this, we find that very few events pass our event selection, making their direct study difficult.

The prompt diphoton sample is generated with `MADGRAPH5_aMC@NLO` at NLO including Born ($q\bar{q} \rightarrow \gamma\gamma$) and Box ($gg \rightarrow \gamma\gamma$) production modes. As with the QCD samples, these are generated with phase space cuts that maximize kinematic overlap with Higgs resonant production. This is then interfaced with `Pythia8` for parton matching and hadronization.

The $\mathrm{H} \rightarrow \gamma\gamma$ background sample accounts for the dominant gluon-fusion production mode ($\mathrm{gg} \rightarrow \mathrm{H}$) only, generated with `MADGRAPH5_aMC@NLO` at next-to-leading order (NLO) with FxFx merging. A systematic is assigned to account for the difference in normalization from the full inclusive Higgs production cross-section.

Drell-Yan ($\mathrm{Z} \rightarrow \mathrm{e}^+\mathrm{e}^-$) samples, though not a significant background source for our signal, are additionally generated as they are used to derive systematics for the m_a regressor (see Sect. 8.3). These are generated with `POWHEG` at NLO, and likewise interfaced with `Pythia8` for parton matching and hadronization.

Table 5.6 Simulated background samples, Run 2018

Data set name
/GluGluHToGG_M125_TuneCP5_13TeV-amcatnloFXFX-pythia8/
/DYToEE_M-50_NNPDF31_TuneCP5_13TeV-powheg-pythia8/
RunIIAutumn18MiniAOD-102X_upgrade2018_realistic_v15-v1/MINIAODSIM
RunIIAutumn18DRPremix-102X_upgrade2018_realistic_v15-v1/AODSIM

For all of the above background samples, the PDFs, tune, detector response, and pileup simulation are as described earlier for the signal samples.

The full list of simulated background samples is provided in Tables 5.4, 5.5 and 5.6.

References

1. Reconstruction of decays to merged photons using end-to-end deep learning with domain continuation in the CMS detector (2022)
2. Curtin D, Essig R, Gori S, Jaiswal P, Katz A, Liu T, Liu Z, McKeen D, Shelton J, Strassler M (2014) Exotic decays of the 125 GeV Higgs boson. Phys Rev D 90
3. Curtin D, Essig R, Gori S, Shelton J (2015) Illuminating dark photons with high-energy colliders.J High Energy Phys 157

Chapter 6
Signal Selection

In this chapter, we describe the selection workflow used for obtaining $\mathrm{H} \rightarrow \mathrm{aa} \rightarrow 4\gamma$ candidate events from the CMS data sets described in the previous chapter. Since the focus of this analysis is the merged $\mathrm{a} \rightarrow \gamma\gamma$ regime, for which the $\mathrm{H} \rightarrow \mathrm{aa} \rightarrow 4\gamma$ event topology mimics that of the SM $\mathrm{H} \rightarrow \gamma\gamma$ decay, the selection criteria closely resemble those of the $\mathrm{H} \rightarrow \gamma\gamma$ analysis. The first part of the selection workflow is the HLT trigger, and is presented in Sect. 6.1. The second part of the selection workflow is the identification of $\mathrm{a} \rightarrow \gamma\gamma$ candidates and is intended to capitalize on the unique features of the $\mathrm{a} \rightarrow \gamma\gamma$ decay topology. This section is divided into two segments: the first segment, given in Sect. 6.2, describes the reconstructed photon preselection, which seeks to emulate the criteria used in the HLT trigger, in terms of standard "offline" variables. The second segment, the $\mathrm{a} \rightarrow \gamma\gamma$ identification in Sect. 6.3, describes the criteria targeting the unique aspects of the $\mathrm{a} \rightarrow \gamma\gamma$ topology, in order to reduce contamination from background objects, namely, single photons and electromagnetically decaying jets. The final part of the selection workflow, presented in Sect. 6.4, details the event selection criteria that leverage the fact that the particles a are produced from Higgs bosons. Since the Higgs boson is itself produced resonantly, the invariant mass of the selected $\mathrm{a} \rightarrow \gamma\gamma$ candidates can be used to further reduce contributions from non-resonant background processes. The chapter concludes with Sect. 6.4 which provides an accounting of the event yields passing the above selection criteria, as well as some rough estimates of the signal sensitivity that can be expected based on these.

In the way of object selection, the $\mathrm{a} \rightarrow \gamma\gamma$ shower pattern, depending on its boost, is subtly differentiated from that of true photons or of electromagnetically-enriched QCD jets faking photons. However, as we are primarily interested in probing the $\mathrm{H} \rightarrow \gamma\gamma$-like phase space for hints of a buried signal, we do not explicitly suppress photons except through the action of the m_Γ regressor.

M. Andrews, *Search for Exotic Higgs Boson Decays to Merged Diphotons*,
Springer Theses, https://doi.org/10.1007/978-3-031-25091-0_6

On the other hand, QCD jets, even if electromangentically-enriched through $\pi^0 \rightarrow \gamma\gamma$ or $\eta \rightarrow \gamma\gamma$ decays, exhibit a distinctly hadronic signature–nearby charged tracks or HCAL deposits from other, hadronic constituents in the jet–that can be exploited for jet-specific background suppression.

6.1 Trigger

The first level of event selection applied at the analysis level is the HLT trigger filter, as discussed in Sect. 2.4. This is an *online* selection applied in real-time at the moment of data taking, as opposed to the event and object selections which are applied *offline*, post-data taking. We require events recorded in the CMS data sets to have fired a diphoton HLT trigger with invariant mass near the Higgs window.

To control QCD jet background contamination which have orders-of-magnitude larger cross-section, the triggering photons–essentially ECAL deposits–are required to pass shower shape and isolation criteria. To control the trigger rate, lower bounds are placed on the transverse energy E_T of the triggering photons, specifically the subleading E_T photon candidate, which may change with data taking period as the LHC luminosity increases. These are:

- 2016:
  `HLT_Diphoton30_18_R9Id_OR_IsoCaloId_AND_HE_R9Id_Mass90_v*`
- 2017, 2018:
  `HLT_Diphoton30_22_R9Id_OR_IsoCaloId_AND_HE_R9Id_Mass90_v*`,
  `HLT_Diphoton30_22_R9Id_OR_IsoCaloId_AND_HE_R9Id_Mass95_v*`[1]

These are identical to those used by the SM $H \rightarrow \gamma\gamma$ analysis [1]. In words, the diphoton trigger objects must have an invariant mass of $m_{\Gamma\Gamma} > 90$ GeV and leading and subleading $E_T > 30$ and 18(22) GeV, respectively for 2016 (2017–2018). In addition, each triggering photon must pass a loose set of ECAL shower shape and isolation criteria. As discussed in the following subsection, these criteria are identical to those applied offline for photon identification. A detailed explanation of these is thus deferred therewith.

A large fraction of the signal selection efficiency lost from the trigger (see Table 6.2) is due to the E_T requirements, which, unfortunately, is constrained by the trigger rate. We find the shower shape requirements of this trigger, however, to be reasonably inclusive: for the lower half of the m_a range, the $a \rightarrow \gamma\gamma$ shower shapes are, to good approximation, identical to that of true photons, with the trigger

[1] The second trigger for 2017/2018, differs from the first only in invariant mass ($m_{\Gamma\Gamma} > 95$ vs 90 GeV) and is, in fact, a subset of the first. It was implemented as a precaution in the event the trigger rate in the first became unmanageable, which eventually did not come to pass. It is included here for consistency with the $H \rightarrow \gamma\gamma$ analysis.

efficiency at signal masses $m_a \approx 400$ MeV more than 95% of the efficiency of signal masses at $m_a \approx 100$ MeV. For reference, the photon identification (ID) efficiency of this trigger for true photons is > 99%[2]. This then drops to about 62% of the true photon ID efficiency for signal masses $m_a \approx 1$ GeV.

6.2 Photon Preselection

At lower masses $m_a \approx 100$ MeV, $a \to \gamma\gamma$ decays closely resemble true photons in both ECAL shower shape and isolation characteristics, while at higher masses $m_a \approx 1$ GeV, they resemble the neutral hadron decays (e.g. $\pi^0 \to \gamma\gamma$) found in electromagnetically-enriched (EM) QCD jets, at least in shower shape. Since EM jets are the primary background considered in the specification of photon criteria in the diphoton triggers, this invariably leads to some loss in signal selection efficiency at higher m_a.

To maximize signal efficiency over the full range of considered m_a, we first apply as loose a photon preselection criteria as is allowed by the photon triggers, and follow this with $a \to \gamma\gamma$-optimized identification criteria. The first part, photon preselection, emulates the photon trigger criteria in terms of the analogous offline reconstructed quantities, and is described in this subsection. The second part, $a \to \gamma\gamma$ identification, is described in the following subsection. For context, both criteria are looser than the EGM "loose" cut-based ID working point.

The photon preselection criteria are listed below.

- Ratio of the energy deposited in the 3×3 crystal window surrounding the most energetic (seed) crystal over that deposited over the entire photon supercluster, $R_9 > 0.5$.
- Ratio of energy deposited in the closest HCAL tower over that deposited within the 5×5 ECAL crystal window of the seed crystal, $H/E < 0.04596$.[3]
- No track seed deposited in pixel tracker within a cone of $\Delta R < 0.3$ of photon position.
- Square root of the covariance of the photon shower shape in the ECAL along the η direction, $\sigma_{i\eta i\eta}$, as in Table 6.1.
- Sum of ECAL energy within a cone of $\Delta R < 0.3$ of the photon position, $\mathcal{I}_\gamma$, as in Table 6.1.
- Sum of p_T of tracks within a cone $\Delta R < 0.3$ of the photon position with the central $\Delta R < 0.03$ excluded, or track isolation, $\mathcal{I}_{tk}$, as in Table 6.1.

The above photon preselection criteria are similar to those used by the $H \to \gamma\gamma$ analysis.

---

[2] In fact, this is measured for electrons using the tag-and-probe method on $Z \to e^+e^-$ events.

[3] This is the only quantity which differs from the trigger photon criteria to accommodate constraints during the derivation of the m_Γ scale and smearing systematics.

Table 6.1 Photon preselection cut values per R_9 category for $\sigma_{i\eta i\eta}$, $\mathcal{I}_\gamma$, and $\mathcal{I}_{\text{tk}}$. See text for variable descriptions

	$\sigma_{i\eta i\eta}$	$\mathcal{I}_\gamma$	$\mathcal{I}_{\text{tk}}$
$R_9 > 0.85$	–	–	–
$0.5 < R_9 \leq 0.85$	< 0.015	$< 4\,\text{GeV}$	$< 6\,\text{GeV}$

6.3 a $\to \gamma\gamma$ Identification

In this subsection, we describe the application of targeted cuts, applied after photon preselection, to optimize the selection of a $\to \gamma\gamma$ candidates. For simplicity, unless otherwise specified, the a $\to \gamma\gamma$ candidate is simply referred to as *the photon candidate* Γ, with the implicit understanding that it corresponds to a single a $\to \gamma\gamma$ decay candidate.

The photon identification criteria primarily optimizes the selection of a $\to \gamma\gamma$ decays against EM-enriched QCD jet backgrounds. This is done by exploiting the distinctly hadronic nature of jets, and by using multivariate-based tools.

For the first, we utilizes the relative charge isolation, $\mathcal{I}_{\text{tk}}/p_{\text{T},\Gamma}$, as a handle on the additional hadronic activity in the jet other than from the primary electromagnetic decay (e.g. one or more $\pi^0 \to \gamma\gamma$s or $\eta \to \gamma\gamma$).

For the multivariate-based background rejection, a dedicated a $\to \gamma\gamma$ vs. SM tagger would ideally be developed to optimally separate a $\to \gamma\gamma$ from QCD jets. This could be a complementary end-to-end ML a $\to \gamma\gamma$ tagger that fully exploit all relevant detector information about the shower shape and track structure. However, as this analysis represents a first application of end-to-end ML techniques in an analysis, in the interest of simplicity, this is left to future work. For the present analysis, we opt instead to use the photon vs. EM-enriched jet discriminator readily available in the `CMSSW` toolkit, namely the EGM photon ID MVA. While not as optimal as a dedicated a $\to \gamma\gamma$ tagger, it nonetheless provides a significant reduction in SM jet backgrounds at minimal cost to a $\to \gamma\gamma$ selection efficiency (see Table 6.2) if a loose enough cut is used.

To optimize the cut values of the above variables, we perform a grid search in ($\mathcal{I}_{\text{tk}}/p_{\text{T},\Gamma}$, EGM MVA) to maximize the final analysis signal significance (see Sect. 9.1). We found little motivation or benefit in tuning additional photon variables.

Lastly, as described in Chap. 7, when using the end-to-end mass regressor, we restrict its use to the predicted mass range of $m_\Gamma = (0, 1.2)\,\text{GeV}$. This effectively rejects out-of-sample $m_\text{a} > 1.2\,\text{GeV}$ a $\to \gamma\gamma$ decays, and rejects about 50% of true photons–typically unconverted photons.

The final photon identification criteria are listed below:

- Ratio $\mathcal{I}_{\text{tk}}/p_{\text{T},\Gamma}$, or relative track isolation, less than 0.07.
- EGM photon ID MVA (photon vs. EM-enriched jet multivariate classifier score) greater than -0.96. Scores range in the interval (-1:jet-like, 1:photon-like).

- Reconstructed particle mass within the window $m_\Gamma = (0, 1.2)$ GeV. This rejects out-of-sample $m_a > 1.2$ GeV a $\to \gamma\gamma$ decays, rejects about 50% of true photons–typically unconverted photons, as explained in Chap. 7.

The first two variables together reject nearly 80% of background events passing photon preselection while preserving 73–90% of signal events.

6.4 Event Selection

To capitalize on the resonant topology of Higgs boson production, the H $\to$ aa $\to 4\gamma$ event selection, similar to the SM H $\to \gamma\gamma$ event selection, applies a number of kinematic requirements on events passing the diphoton trigger.

In order to maintain orthogonality with complementary H $\to$ aa $\to 4\gamma$ analyses at CMS that seek to probe resolved decay regimes, we impose requirements on the number of reconstructed photons[4] and the number of these passing photon identification criteria. In particular, this analysis focuses on the phase space containing exactly two photons passing photon identification. This is the so-called "merged-merged" regime where both legs of the H $\to$ aa $\to 4\gamma$ decay are reconstructed as single photon candidates. Other efforts ongoing at CMS include a potential "merged-resolved" analysis, where only one of the a $\to \gamma\gamma$ decays is resolved, and the "resolved-resolved" analysis [2], where both decays are resolved. Our chosen mass range has minimal overlap with these other phase spaces.

To simplify the scope of the end-to-end m_Γ regression technique, we further restrict the analysis to events where both of the photons passing photon identification—the photons whose mass m_Γ are to be regressed—deposit in the ECAL barrel. This corresponds to the "EB-EB" category found in the SM H $\to \gamma\gamma$ analysis. A follow-up version of the m_Γ regressor covering the endcaps, and potentially involving the much finer-grained ECAL Preshower, is being investigated for a future analysis.

The full list of event selection criteria is given below:

- Events passing the diphoton trigger are required to contain no more than three reconstructed photons. A third photon is allowed, since, at $m_a \approx 1$ GeV, a non-negligible fraction of a $\to \gamma\gamma$ decays are resolved as two distinct photon candidates. Allowing a third photon, however, generally improves signal efficiency across the m_a range as it accommodates signal events that may include a jet faking a photon from either pileup or from associated Higgs production. This allows more than 90% of signal events passing the trigger to be selected.
- Selected events must contain exactly two barrel photons that pass photon identification criteria with pseudorapidity $|\eta| < 1.4$. These are the two photons whose masses m_Γ will be regressed to construct the 2D-m_Γ distribution. The third photon, if present, must not pass photon identification criteria—in which case it would be

[4] As present in the `pat::photon` (MINIAOD) collection.

categorized as a "merged-resolved" event—but is allowed to deposit in the endcaps, upto $|\eta| < 2.5$. Events with a pair of reconstructed photons in the barrel still account for two-thirds of events with two or three reconstructed photons. The condition on the third photon preserves between 72% ($m_{\mathrm{a}} \approx 1$ GeV) to 98% ($m_{\mathrm{a}} \approx 100$ MeV) of the yield from signal events with at least two EB photons passing photon identification.

- The invariant mass of the diphoton system $m_{\Gamma\Gamma}$, where each Γ denotes the single, selected, merged photon candidate passing photon identification, must fall within the Higgs mass window $100 < m_{\Gamma\Gamma} < 180$ GeV. Of note, with increasing m_{a}, the measured energy of the merged photon candidate becomes increasingly underestimated by standard particle reconstruction techniques. This, in turn, skews and distorts the reconstructed Higgs peak downward. While this effect was used in [3] to indirectly probe the presence of $\mathrm{H} \rightarrow \mathrm{aa} \rightarrow 4\gamma$ events, it adversely affects signal selection efficiency. It is a major limiting factor in extending the analyzed m_{a} range higher but also presents a prime use case for end-to-end deep learning-based energy reconstruction.
- To prevent the sculpting of the mass distributions from the turn-on of the p_{T} thresholds in the trigger, the selected diphotons are required to have a scaled transverse momentum of $p_{\mathrm{T},\Gamma}/m_{\Gamma\Gamma} > \frac{1}{3}, \frac{1}{4}$ for the leading and subleading selected photon candidate, respectively. The cut values are chosen for consistency with the SM $\mathrm{H} \rightarrow \gamma\gamma$ analysis.

No further multivariate event classification is performed in this analysis even though it is done for the SM $\mathrm{H} \rightarrow \gamma\gamma$ analysis. This is left to future work as well.

6.5 Event Yields

The expected event yields in data (assumed to be all background) and in a hypothetical $\mathrm{H} \rightarrow \mathrm{aa} \rightarrow 4\gamma$ signal using the above event selection criteria is summarized in Table 6.2. We estimate a signal selection efficiency of between 4–16%, depending on m_{a}, at a background rejection rate of > 99%.

Choosing, as a matter of convention, a hypothetical signal cross-section of $\sigma(\mathrm{H} \rightarrow \mathrm{aa} \rightarrow 4\gamma) = 1$ pb, we can compare the estimated signal sensitivity between the different signal mass points. As a rough proxy for signal significance, we calculate the quantity $S/\sqrt{B}$, where $S(B)$ is the expected signal (background) yield. This is motivated by the fact that, under Poisson statistics, one standard deviation of statistical uncertainty in the background model, scales as $\sqrt{B}$. We find a value of $S/\sqrt{B} \approx 7$ to 25, depending on m_{a}. Of course, these simple significance estimates are based on arbitrary signal cross-sections and do not take into account the detailed shapes of the distributions in the 2D-m_{Γ} plane, which can alter the final signal sensitivity (see Sect. 9.1) considerably. More realistic signal cross-sections are discussed in Sect. 8.1 (Table 6.3).

Table 6.2 Absolute (N_{events}) and relative (f_{events}) event yields for the ma-SR signal region

Criteria	$\mathbf{N}_{\text{events}}$	$\mathbf{f}_{\text{events}}$		$\mathbf{f}_{\text{events}}\,(\mathbf{m}_{\text{a}})$	
	Data	Data	100 MeV	400 MeV	1 GeV
None	2,156,696,722	1.000	1.000	1.000	1.000
Trigger	316,447,814	0.147	0.578	0.547	0.388
N_{Γ}	292,164,520	0.135	0.563	0.532	0.359
Preselection	29,189,279	0.014	0.269	0.254	0.141
$m_{\Gamma\Gamma}$	13,865,435	0.006	0.265	0.248	0.110
$p_{\text{T},\Gamma}/m_{\Gamma\Gamma}$	9,920,283	0.005	0.249	0.237	0.098
EGM MVA	5,339,303	0.002	0.249	0.222	0.087
$\mathcal{I}_{\text{tk}}/p_{\text{T},\Gamma}$	2,024,716	0.001	0.236	0.214	0.083
m_{Γ} window	1,097,246	0.001	0.129	0.185	0.061

Table 6.3 Estimated signal sensitivity for a signal with hypothetical cross section $\sigma(\text{H} \rightarrow \text{aa} \rightarrow 4\gamma) = 1$ pb. Note that these do not take into account the shape of the 2D-m_{Γ} distribution which significantly alters the final sensitivity

m_{a}	100 MeV	400 MeV	1 GeV
N_{events}	17661	25244	8348
$S/\sqrt{B}$	16.9	24.1	8.0
$S/(S+B)$	0.016	0.022	0.008

References

1. Khachatryan V et al (2014) Observation of the diphoton decay of the Higgs boson and measurement of its properties. Eur Phys J C 74(10):3076
2. Search for the exotic decay of the Higgs boson into two light pseudoscalars with four photons in the final state in proton-proton collisions at $\sqrt{s} = 13$ TeV (2022)
3. Collaboration A (2012) Search for a Higgs boson decaying to four photons through light CP-odd scalar coupling using 4.9 fb^{-1} of 7 TeV pp collision data taken with ATLAS detector at the LHC 7

Chapter 7
$a \to \gamma\gamma$ Mass Regression

In this chapter, we describe the algorithm for reconstructing the invariant mass of the $a \to \gamma\gamma$ decay, using the end-to-end deep learning (ML) technique. This algorithm, the m_Γ regressor, and the ideas that have made it possible, represents the culmination of years of prior work, both published [1–4] and unpublished, investigating the broader feasibility of using ML on very low-level collider data, to achieve breakthrough reconstruction performance. In Sect. 7.1, we describe the construction of the detector images used as inputs to the ML algorithm. In this first application of the end-to-end technique, only data from the ECAL barrel is used. Then, in Sect. 7.2, we describe the ML training strategy for the m_Γ regressor, including the domain continuation technique developed to address the challenges of the ultra-boosted regime. A validation of the m_Γ regressor then follows in three parts. In the first part, in Sect. 7.3, we present a basic evaluation of the ML training using a "test set" of simulated samples closely resembling the those used to train the m_Γ regressor. Then, in Sect. 7.4, a more realistic sample of simulated $H \to aa \to 4\gamma$ events, as well as actual recorded data, is used to more thoroughly benchmark, and illustrate the potential of, the m_Γ regressor against close alternatives. Finally, in Sect. 7.5 we conduct a series of $H \to aa \to 4\gamma$-specific studies, to asses how the m_Γ performs under different kinematic regimes and detector conditions, as well as comparing the response in data versus simulation. Additional details about the m_Γ regressor can be found in the dedicated paper in [4].

7.1 Image Construction

Since the photons from $a \to \gamma\gamma$ decays deposit energy primarily in the ECAL, for simplicity, we use an image construction strategy that only consists of ECAL information. We take a 32×32 matrix of ECAL crystals around the most energetic (seed) crystal of the reconstructed photon candidate and create an image array. This corresponds to an angular cone of $\sqrt{\Delta\eta^2 + \Delta\phi^2} \approx 0.3$, and ensures the subleading photon in the $a \to \gamma\gamma$ decay is fully contained in the image. Each pixel in the

M. Andrews, *Search for Exotic Higgs Boson Decays to Merged Diphotons*,
Springer Theses, https://doi.org/10.1007/978-3-031-25091-0_7

image exactly corresponds to the energy deposited in a single ECAL crystal. These energy depositions represent the actual interaction of the incident photon with the detector material (c.f. Fig. 3.12, right column). This approach is distinct from those that use PF candidates, in which case the a $\rightarrow \gamma\gamma$ decay would always appear as a single image pixel. No rotation is performed on the images as this is, in general, a destructive operation that distorts the particle shower pattern. For simplicity, only photons depositing energy in the barrel section of the ECAL are used in this paper. For the case of general particle decays, the ECAL images can be combined with additional subdetector images [3], or parsed into a multi-subdetector graph if one is using such an architecture. In particular, incorporating tracking detector images, even for the present a $\rightarrow \gamma\gamma$ images, may enable better accounting of contributions from e^+e^+ conversions or pileup, but are not yet considered in this study.

As noted in Sect. 2, because of the η-dependent material budget of the inner tracker, electromagnetic shower development varies significantly with η. Since the (η, ϕ) position of the shower is lost after the image cropping procedure, we perform a number of augmentations to recover this information during the training of the ML algorithm. The first is to split the ECAL images described above into a two-layer image that instead contains the transverse and longitudinal components of the crystal energy. The transverse and longitudinal crystal energies are defined as $E_\mathrm{T} = E \sin(\theta)$ and $E_\mathrm{Z} = E \cos(\theta)$, respectively, where E is the crystal energy and θ is the polar angle of the energy deposit. The second image augmentation is to include the crystal seed coordinates. The energy decomposition is found to be more beneficial.

The ECAL detector data exists at various levels of processing: from minimally processed detector-level data to the filtered and clustered version more easily accessible at the analysis level. As the clustered detector data is historically optimized for the needs of PF, for a new approach such as end-to-end ML, it is worth revisiting this assumption. As discussed in Appendix A.2, training on minimally processed instead of clustered data significantly improves our final results. We thus emphasize our earlier statement that using minimally processed detector data is critical to realizing the full potential of ML. We make exclusive use of minimally processed detector data in this paper and note that accessing such data is becoming a logistical challenge because of the trend toward more compact CMS file formats, necessitated by the growing volume of LHC data.

7.2 Training

For training, we use a sample of 780k simulated a $\rightarrow \gamma\gamma$ decays in which particle a has a narrow width and negligible lifetime. The sample has an ensemble of *continuously distributed* masses m_a, with $p_\mathrm{T,a}^\mathrm{gen} = 20$–$100$ GeV, $m_\mathrm{a} = 0$–1.6 GeV, and $|\eta_\mathrm{a}| < 1.4$, corresponding to Lorentz boosts of $\gamma_L \sim 10^1$ to 10^3 for m_a from 100 MeV to 1 GeV, respectively. The phase space of a $\rightarrow \gamma\gamma$ decays is chosen such that samples passing the photon identification criteria are uniformly distributed in $(p_{T,a}^{gen}, m_a)$, so as not to bias the training. The simulated samples account for pileup and use 2017 detector conditions only.

These samples are then fed to a ResNet CNN-based mass regression algorithm[1] Each image associated with a reconstructed photon candidate Γ is then passed to the ResNet model, which outputs to a global maximum pooling layer representing the feature vector learned by the CNN. These outputs are then concatenated with the crystal seed coordinates of the photon candidate. The concatenated vector is then fully connected to a final output node that represents the predicted or regressed particle mass for that photon candidate.

To train the mass regressor, the predicted mass m_Γ is compared to the true (generated) particle mass m_a by calculating the absolute error loss function $|m_\Gamma - m_a|$, averaged over the training batch. Other loss functions are found to be equally performant. This loss function is then minimized using the `ADAM` optimizer [5]. This procedure represents our basic training strategy.

The $m_a \to 0$ boundary problem. Relying solely on this basic training strategy has significant performance limitations. As shown in the left plot of Fig. 7.1, naively training the mass regressor as described above results in a nonlinear response near either boundary of the mass regression range. At the high-m_a boundary, this issue can be resolved by trivially extending the training mass range to $m_a \approx 1.6$ GeV and removing the extended mass range during inference. Equivalently, one can simply limit the usable upper range of the mass regressor to $m_\Gamma < 1.2$ GeV.

This approach cannot, however, be used in any obvious way for the low-m_a boundary, since it is constrained by the physical requirement $m_a > 0$ GeV. The mass region $m_\Gamma \lesssim 200$ MeV, of considerable theoretical interest for the diphoton decay mode in BSM models (see Sect. 3.3.1), would therefore be inaccessible. The use of the mass regressor as a tool for reconstructing $\pi^0 \to \gamma\gamma$ decays would also be lost. Moreover, significant biases in the reconstructed masses of true photons would also arise. As illustrated in the right plot of Fig. 7.1, photons would be reconstructed as a peak around $m_\Gamma \lesssim 200$ MeV, reducing even further the usable range of the mass regressor when photon backgrounds are taken into account.

Domain continuation to negative masses. While every detector in which one is trying to reconstruct a particle mass has a finite resolution $\sigma(m)$, our attempts to distinguish discrete mass points in the low-m_a regime, as summarized in Table 7.1, suggest discriminating power is available even at the limit of the detector resolution. It is thus worthwhile to understand why the naive mass regressor is unable to exploit this region.

Fundamentally, the boundary problem arises because, when training the mass regressor, the physically observable $a \to \gamma\gamma$ invariant mass distribution becomes under-represented for samples with $m_a < \sigma(m_a)$. This issue is illustrated in the left

[1] While other architectures exist, including those that use graph-based architectures, emphasis in this analysis was placed on the general reconstruction technique rather than ML model optimization. Nonetheless, we expect the ResNet architecture to deliver close to optimal performance.

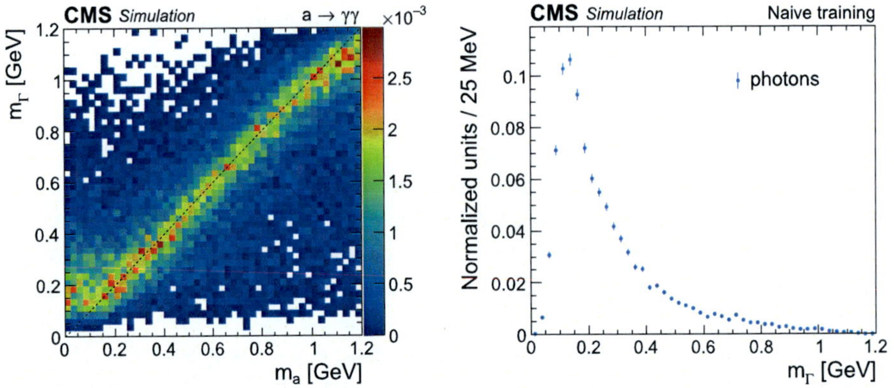

Fig. 7.1 Left plot: the reconstructed naively trained mass regressor value m_Γ versus the generated m_a value for simulated $a \to \gamma\gamma$ decays generated uniformly in (p_T, m_a). The reconstructed m_Γ is normalized in vertical slices of the generated m_a. Right plot: the reconstructed m_Γ distribution for simulated single-photon samples, resulting in a distinct peak in the low-m_Γ region. Credit: Ref. [4]

Table 7.1 Areas-under-the curve (AUC) of the Receiver Operating Characteristic (ROC) for end-to-end γ versus $a \to \gamma\gamma$ ECAL shower classification. Events taken from simulated H and $H \to aa \to 4\gamma$ samples, respectively. Trained on approximately 200k events with balanced class proportions using a similar network architecture to that described in the text for $a \to \gamma\gamma$ mass regression

γ versus $a \to \gamma\gamma$ Classification m_a	ROC AUC
100 MeV	0.74
200 MeV	0.88
400 MeV	0.96

plot of Fig. 7.2. For samples $m_a \approx \sigma(m_a)$, the full, observable mass distribution (f_{obs}), illustrated as a Gaussian distribution, is barely represented in the training set. As $m_a \to 0$, shown in the middle plot of Fig. 7.2, only half of the mass distribution is now observable. For these under-represented samples, the behavior of the mass regressor is to default to the last full mass distribution at $m_a \approx \sigma(m_a)$, causing the gap and accumulation of masses at $m_\Gamma \approx 200$ MeV. More generally, this boundary problem manifests itself when regressing a quantity q, with resolution $\sigma(q)$, over the range (a, b), for samples with $q \lesssim a + \sigma(q)$ or $q \gtrsim b - \sigma(q)$. Only in the limit $\sigma(q) \ll a, b$ does this effect become negligible at either boundary.

This motivates a solution for the low-m_a boundary problem by extending the regression range beyond $m_a = 0$, into the nonphysical domain, and populating it with "topologically similar" samples. We thus augment the training set with samples artificially and randomly labeled with negative masses. During inference, we remove the nonphysical predictions $m_\Gamma < 0$. As a topologically similar sample, either a sample of fixed mass $m_a \approx 0.001$ MeV decays or true photons can be used. In this paper,

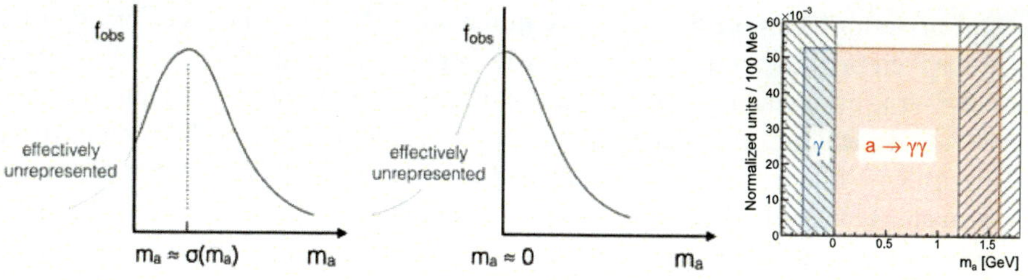

Fig. 7.2 Pictorial representation of the $m_a \to 0$ boundary problem occurring when attempting to regress below the mass resolution. Left plot: The distribution of observable a $\to \gamma\gamma$ invariant masses (f_{obs}) versus the generated m_a. When $m_a \approx \sigma(m_a)$, the left tail of the mass distribution becomes under-represented in the training set. Middle plot: As $m_a \to 0$, only half of the mass distribution is represented. The regressor subsequently defaults to the last full mass distribution at $m_a \approx \sigma(m_a)$. Right plot: With domain continuation, the original training samples (a $\to \gamma\gamma$, red region) are augmented with topologically similar samples that are randomly assigned nonphysical masses (γ, blue region). This allows the regressor to see a full mass distribution over the entire region of interest (nonhatched region). Predictions outside of these (hatched regions) are not used. Credit: Ref. [4]

we use the latter, although we find either works well. If we require that the "negative mass" samples have the same mass density as the "positive mass" in the training set (c.f. Fig. 7.2, right plot), then only a single hyper-parameter is needed, the minimum artificial mass value, labeled $\min(m_a)$. This can be set by choosing the least-negative value that closes the low-m_a gap (c.f. Fig. 7.1, left plot) and linearizes the mass response in the physical domain, $m_\Gamma > 0$. We find a value of $\min(m_a) = -300$ MeV to be sufficient. Other applications may seek to optimize both the minimum artificial mass value and the number density of the augmented samples. Note that having the augmented samples carry negative mass values is simply an accident of the low-m_a boundary coinciding with $m_\Gamma = 0$. Had the boundary coincided with a positive mass, positive artificial mass values would be involved as well.

The above procedure effectively tricks the mass regressor into seeing a full invariant mass distribution for all physical a $\to \gamma\gamma$ decays, even when they reside below the detector mass resolution. As a result, the full low-m_a regime becomes accessible. In addition, this procedure provides a simple way for suppressing photon backgrounds. Since true photons will tend to be predicted with negative masses, removing samples with $m_\Gamma < 0$ reduces photon contributions in a mass decorrelated way. The only trade-off is that low-m_a samples incur a selection efficiency to be regressed within $m_\Gamma > 0$. However, this is expected for most a $\to \gamma\gamma$ merged cases that cannot be distinguished from true photons.

By analogy to complex analysis, we denote the above procedure as *domain continuation*. Similar procedures, however, can also be found in the statistical tests used in high energy physics [6]. Our final training strategy implements domain continuation on top of the basic training strategy described at the beginning of this section.

Out-of-sample response. An important feature of ML-based regression algorithms is that their predictions are bound by the regression range on which they were trained. This is true even when out-of-sample candidates are presented to the mass regressor, potentially leading to unexpected peaks and other features in the predicted mass spectrum. While hadronic jets are indeed out-of-sample, it is desirable not to reject them at the stage of the mass regressor in order to enable the reconstruction of, e.g., embedded $\pi^0 \to \gamma\gamma$ decays. If desired, these can instead be suppressed by altering the reconstructed photon selection criteria.

On the other hand, $\mathrm{a} \to \gamma\gamma$ decays from more massive a particles than were used during training can potentially be regressed as a false mass peak near the upper m_a boundary. For this and other reasons stated earlier, when addressing the boundary problem at the upper mass range, we ignore predictions above $m_\Gamma > 1.2$ GeV. Lastly, to suppress photons, as was already noted earlier, we ignore predictions with $m_\Gamma < 0$. During inference, the use of the mass regressor is thus limited to samples predicted within the region of interest (ROI): $m_\mathrm{a} - \mathrm{ROI} \in [0, 1.2]$ GeV. The impact of this method on the sample selection efficiency is estimated in the following section.

7.3 Validation

To validate the training of the mass regressor, and to characterize its performance, we use a sample of $\mathrm{a} \to \gamma\gamma$ particle decays with continuous mass m_a. A statistically independent test set of 26k $\mathrm{a} \to \gamma\gamma$ decays is used to perform these validations. The predicted versus generated mass is shown in the upper-left plot of Fig. 7.3. We observe a linear and well-behaved mass response throughout the full range of masses in the $m_\mathrm{a} - \mathrm{ROI}$. In particular, the mass regressor is able to probe the low-m_a regime for which it exhibits a gentle and gradual loss in resolution upon approaching the $m_\Gamma = 0$ boundary. This performance confirms the ability of the end-to-end ML technique to access the highest boost regimes, where shower and instrumental merging are present, yet maintain performance into the high-m_a regime, where the particle showers become resolved. The predicted m_Γ distribution (blue points) is found to be consistent with the generated m_a one (red band), within statistical uncertainties, in the $m_\mathrm{a} - \mathrm{ROI}$ (non-hatched region).

To approximate the absolute and relative mass resolution, we calculate the mean absolute error $\mathrm{MAE} = \langle m_\mathrm{a} - m_\Gamma \rangle$, and the mean relative error $\mathrm{MRE} = \langle (m_\mathrm{a} - m_\Gamma)/m_\mathrm{a} \rangle$, respectively. The lower-left plot of Fig. 7.3 displays the MAE (blue circles) and MRE (red squares) as functions of the generated mass. The MAE varies between 0.13–0.2 GeV for m_a in the range 0.1–1.2 GeV, corresponding to mean boosts of $\langle \gamma_L \rangle = 600$–50, respectively. In general, the mass resolution worsens with increasing m_a, as reflected in the MAE trend. However, the relative mass resolution tends to improve with mass, as is evident in the MRE distribution, converging to about 20%. For fixed regression efficiency, improved relative resolution implies better signal significance. Notably, for $m_\mathrm{a} \lesssim 0.3$ GeV, the MAE starts worsening with decreasing mass. This can be attributed to the gradual deterioration of the mass regressor below the detector's mass resolution.

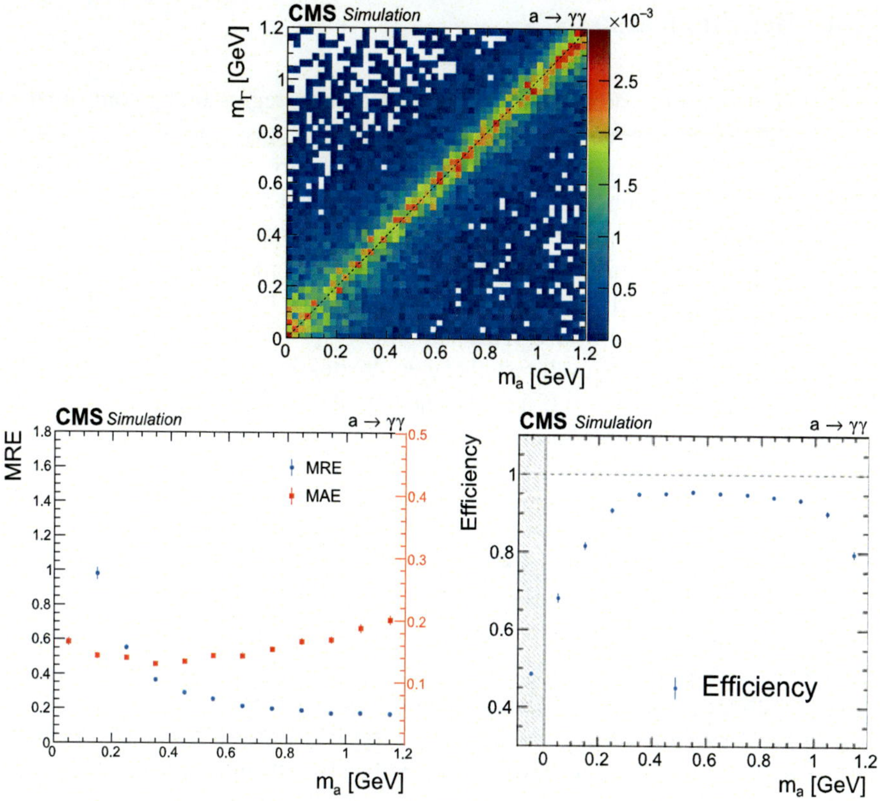

Fig. 7.3 Mass regression performance for simulated a $\rightarrow \gamma\gamma$ samples generated uniformly in $(p_{\mathrm{T}}, m_{\mathrm{a}})$, corresponding to mean boosts approximately in the range $\langle \gamma_L \rangle = 600$–$50$ for $m_{\mathrm{a}} = 0.1$–1.2 GeV, respectively. Upper: Predicted m_Γ versus generated m_{a}. The predicted m_Γ is normalized in vertical slices of the generated m_{a}. Lower left: The MAE (blue circles, use left scale) and MRE (red squares, use right scale) versus the generated m_{a}. For clarity, the MRE for $m_{\mathrm{a}} < 0.1$ GeV is suppressed since its value diverges as $m_{\mathrm{a}} \rightarrow 0$. Lower right: The m_{a} regression efficiency as a function of the generated m_{a}. The hatched region shows the efficiency for true photons. The vertical bars on the points show the statistical uncertainty in the simulated sample. Credit: Ref. [4]

The above figures-of-merit are achieved with a regression efficiency between 70–95%, as shown in the lower-right plot of Fig. 7.3. The regression efficiency is defined as the fraction of a sample in a given m_{a} bin, to have m_Γ within $m_{\mathrm{a}} -$ ROI. For a fixed mass resolution, a higher regression efficiency roughly translates to better signal significance. The efficiency is primarily driven by how much of the mass peak can fit within $m_{\mathrm{a}} -$ ROI. Thus, it is highest at the midway point of $m_{\mathrm{a}} -$ ROI and falls off to either side. The relatively poorer mass reconstruction at low m_{a} causes the efficiency to fall off more steeply for the former. About 50% of true photons are rejected by the $m_{\mathrm{a}} -$ ROI requirement, as seen in the hatched region of the lower-right plot in Fig. 7.3. Photons with $m_\Gamma > 0$ are primarily due to $\mathrm{e}^+\mathrm{e}^+$ conversions.

7.4 Benchmarking

To benchmark the performance of the *end-to-end* mass regressor, we compare it with two traditional reconstruction strategies: a *photon NN*-based mass regressor trained on shower-shape and isolation variables, and a *3 × 3* shower clustering algorithm [7]. The photon NN is trained on a mix of 11 engineered shower-shape and isolation variables, identical to those used by CMS for multivariate photon tagging [8], using a fully-connected neural network. For an even comparison, it is also trained with domain continuation. The 3×3 algorithm is similar to that used for low-energy $\pi^0 \to \gamma\gamma$ reconstruction for ECAL calibration [7]. It first identifies local maxima (seeds) above some energy threshold, then creates 3×3 crystal matrices around these seeds to find clusters. If a pair of nearby clusters is found, the reconstructed mass is calculated as the invariant mass of the two clusters.

7.4.1 $a \to \gamma\gamma$ in Simulated $H \to aa \to 4\gamma$

Benchmarking on simulated data allows the mass spectra for $\mathrm{a} \to \gamma\gamma$ decays for different fixed-mass values to be compared. As described in Sect. 5, we use $\mathrm{a} \to \gamma\gamma$ samples obtained from simulated $\mathrm{H} \to \mathrm{aa} \to 4\gamma$ events with masses $m_{\mathrm{a}} = 0.1, 0.4$, and 1 GeV. In these events, the particle energy is distributed around a median of $E_{\mathrm{a}} \approx m_{\mathrm{H}}/2 \approx 60\,\mathrm{GeV}$, corresponding to median boosts of $\langle\gamma_L\rangle \approx 600, 150$, and 60 for the respective m_{a} masses. The reconstructed mass spectra are shown in Fig. 7.4 for the different algorithms and mass points. For each mass point, representing a different median boost regime, the samples are further broken down by ranges of reconstructed $p_{\mathrm{T},\Gamma}$, to highlight the stability of the mass spectrum with energy. These $p_{\mathrm{T},\Gamma}$ ranges are:

- Low $p_{\mathrm{T},\Gamma}$: $30 < p_{\mathrm{T},\Gamma} < 55\,\mathrm{GeV}$,
- Mid $p_{\mathrm{T},\Gamma}$: $55 < p_{\mathrm{T},\Gamma} < 70\,\mathrm{GeV}$,
- High $p_{\mathrm{T},\Gamma}$: $70 < p_{\mathrm{T},\Gamma} < 100\,\mathrm{GeV}$,
- Ultra $p_{\mathrm{T},\Gamma}$: $p_{\mathrm{T},\Gamma} > 100\,\mathrm{GeV}$.

Some overlap in the boost is thus expected between different mass points. Recall that the mass regressor has only been trained on samples with $p_{\mathrm{T},\Gamma} < 100\,\mathrm{GeV}$.

For boosts $\langle\gamma_L\rangle \approx 60$ and $m_{\mathrm{a}} = 1\,\mathrm{GeV}$, only the end-to-end (Fig. 7.4, top-left plot) is able to consistently reconstruct the peak for all $p_{\mathrm{T},\Gamma}$ ranges. The position of the mass peak also remains stable, with the resolution improving in the high-$p_{\mathrm{T},\Gamma}$ category. The end-to-end regression performs best when the $\mathrm{a} \to \gamma\gamma$ is moderately merged, and neither fully resolved nor fully merged. The mass peak in the ultra-$p_{\mathrm{T},\Gamma}$ category is well-behaved despite being outside the trained phase space. This demonstrates that the phase space extrapolation is effective for internally learned quantities like $p_{\mathrm{T},\Gamma}$.

Additionally, for $m_{\mathrm{a}} = 1\,\mathrm{GeV}$, the photon NN (Fig. 7.4, top-middle plot) has difficulty reconstructing the mass peak, except at the higher $p_{\mathrm{T},\Gamma}$ categories. This can be understood in terms of the information content of the photon variables on which the algorithm was trained. At higher $p_{\mathrm{T},\Gamma}$, the two photons are more likely to be moderately merged so that their showers are contained within the 5×5 crystal block in which the shower-shape variables are defined. At lower $p_{\mathrm{T},\Gamma}$, the two photons are more often resolved so that the lower-energy photon shower falls outside the 5×5 crystal block. The photon NN must then rely on the isolation variables, which are defined much more coarsely over a cone of $\sqrt{\Delta\eta^2 + \Delta\phi^2} < 0.3$ about the seed crystal. Since these have much less discriminating power, this results in a steep fall-off in reconstruction performance. To improve the performance, the photon NN could be augmented with the momentum components of the lower-energy photon, in instances where the PF is able to reconstruct it.

Lastly, for $m_{\mathrm{a}} = 1$ GeV, the 3×3 algorithm (Fig. 7.4, top-right plot) is the only one competitive with the end-to-end method for lower $p_{\mathrm{T},\Gamma}$. As the photon clusters become resolved, the 3×3 method thus becomes an effective tool for mass reconstruction. However, as soon as the clusters begin to merge at higher $p_{\mathrm{T},\Gamma}$, a sudden drop-off in reconstruction efficiency occurs since the 3×3 algorithm is unable to compute a mass for a single cluster. A spurious peak develops at $m_\Gamma \approx 500$ MeV for decays with sufficient showering prior to the ECAL. The 3×3 method is thus only ideal for a limited range of low boosts. As discussed at the end of the following subsection and shown in Fig. 7.7, the performance of the end-to-end technique at lower boosts can likely be improved by extending the training phase space accordingly.

For boosts $\langle\gamma_L\rangle \approx 150$ and $m_{\mathrm{a}} = 400$ MeV, the end-to-end method (Fig. 7.4, second row, left plot) is able to reconstruct the mass peak with full sensitivity across most of the $p_{\mathrm{T},\Gamma}$ ranges. Only at the highest-$p_{\mathrm{T},\Gamma}$ range does the mass peak significantly degrade, although it is still reasonably well-behaved. Training with higher $p_{\mathrm{T},\Gamma}$ could potentially improve this behavior. The photon NN performs its best in this regime (Fig. 7.4, second row, middle plot) because a majority of the photon showers fall within the 5×5 crystal block. However, the mass resolution is still significantly worse compared to the end-to-end method. The 3×3 algorithm (Fig. 7.4, second row, right plot) is barely able to reconstruct a mass peak for these boosts.

For boosts $\langle\gamma_L\rangle \approx 600$ and $m_{\mathrm{a}} = 100$ MeV, the end-to-end method (Fig. 7.4, third row, left plot) reaches the limits of its sensitivity, although it is still usable. Notably, even at this limit, the position of the mass peak remains stable with $p_{\mathrm{T},\Gamma}$. This is not the case for the photon NN (Fig. 7.4, third row, middle plot) whose peak becomes erratic and displaced with increasing $p_{\mathrm{T},\Gamma}$. The 3×3 method is not able to calculate a mass at this level of merging.

For reference, the predicted mass spectrum for photons is shown in the bottom row of Fig. 7.4. Both the end-to-end (left) and photon NN (middle) are able to regress to the $m_\Gamma \approx 0$ GeV boundary, with a smoothly falling distribution since they were trained with domain continuation (c.f. Fig. 7.1, right plot). The remaining photons within m_{a} − ROI come from photon conversions that acquire an effective mass because of nuclear interactions.

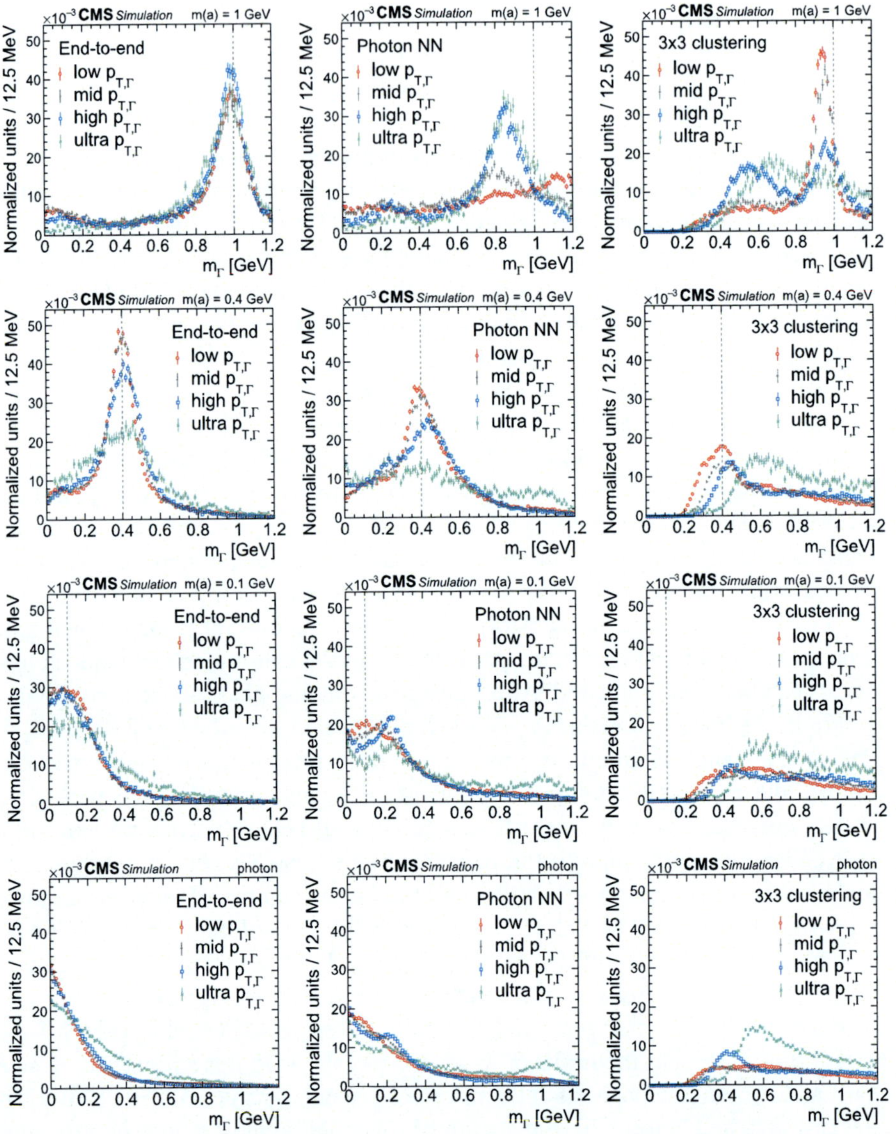

Fig. 7.4 Reconstructed mass spectra for end-to-end (left column), photon NN (middle column), and 3 × 3 algorithm (right column) for $a \rightarrow \gamma\gamma$ decays with $m_a = 1$ GeV (top row), $m_a = 400$ MeV (second row), $m_a = 100$ MeV (third row), and photons (bottom row) generated using an energy distribution with median $E_a \approx 60$ GeV. For each panel, the mass spectra are separated by reconstructed $p_{T,\Gamma}$ value into ranges of $30 < p_{T,\Gamma} < 55$ GeV (red circles, low $p_{T,\Gamma}$), $55 < p_{T,\Gamma} < 70$ GeV (gray + symbol, mid $p_{T,\Gamma}$), $70 < p_{T,\Gamma} < 100$ GeV (blue square, high $p_{T,\Gamma}$), and $p_{T,\Gamma} > 100$ GeV (green × symbol, ultra $p_{T,\Gamma}$). All the mass spectra are normalized to unity. The vertical dotted line shows the input m_a value. Credit: Ref. [4]

7.4.2 $\pi^0 \rightarrow \gamma\gamma$ in Data

To validate the findings from the simulated data above, we perform a benchmark using merged photons from γ + jet events in recorded data. If the jet contains an energetic, collimated neutral meson decay, typically a $\pi^0 \rightarrow \gamma\gamma$ or $\eta \rightarrow \gamma\gamma$, it will be misreconstructed as a single photon Γ and the event will pass a diphoton trigger. Since the energy of the jet will, in general, be shared among several constituent particles, the π^0 is more likely to be reconstructed as the lower-energy photon in the event. A data sample enriched in merged photons is thus obtained by selecting events passing a diphoton trigger and selecting the lower-energy reconstructed photon, which we additionally require to pass our selection criteria. The selected sample is then given to the m_Γ regressor, whose output we study below. We emphasize that the mass regressor is being used to reconstruct the m_Γ of individual reconstructed photons, which we assume to be merged photons, not the invariant mass of the reconstructed diphoton event itself.

However, an important caveat in reconstructing the m_Γ of energetic photons within jets is the presence of other hadrons within the jet. At the energies on which the m_Γ regressor was trained, we emphasize that *it is no longer the case that neutral meson decays are well-isolated in jets*, a main point of distinction compared to the isolated $a \rightarrow \gamma\gamma$ decays used to train the regressor. In general, the neutral meson decay will be collimated with other hadrons, including, potentially, several merged π^0 decays. The effect of these additional hadrons is to smear and distort the resulting m_Γ spectrum and introduce an energy dependence in the m_Γ value, as illustrated in Fig. 7.5.

For the purposes of validating the m_Γ regressor in this section, we therefore restrict our study to events in data passing the lower-mass diphoton trigger[2], $55 < m_{\Gamma\Gamma} < 100$ GeV, $20 < p_{T,\Gamma} < 35$ GeV and require tighter shower-shape criteria, in order to increase the contribution from well-isolated $\pi^0 \rightarrow \gamma\gamma$ decays that more closely resemble the $a \rightarrow \gamma\gamma$ decay. However, note that these tighter criteria only mitigate the stated effects, but the impact of these effects remains visible.

At these energies, the π^0, with a mass of 135 MeV, is boosted to approximately the range $\gamma_L = 150$–250, putting its invariant mass reconstruction out of reach of all but the end-to-end mass regressor. Also present is the η, with a larger mass of 550 MeV, which, though produced with a much lower cross section, is boosted to only about the range $\gamma_L = 30$–60, just within reach of the 3×3 algorithm.

As clearly seen in Fig. 7.6, the end-to-end method (red circles) is able to reconstruct a prominent π^0 peak. Indeed, it is the only algorithm able to do so. The photon NN (blue $\times$ symbol) exhibits an erratic response, suggesting it does not have the information granularity needed to probe this regime. Likewise, the 3×3 method (red $+$ symbol) is unable to reconstruct the π^0 peak at all. It is, however, able to reconstruct the η peak, as expected.

We attribute the weaker η peak in the end-to-end method to the aforementioned smearing effect of additional hadrons in the jet and the advantage of the 3×3 method

---

[2] `HLT_Diphoton30PV_18PV_R9Id_AND_IsoCaloId_AND_HE_R9Id_PixelVeto_Mass55`.

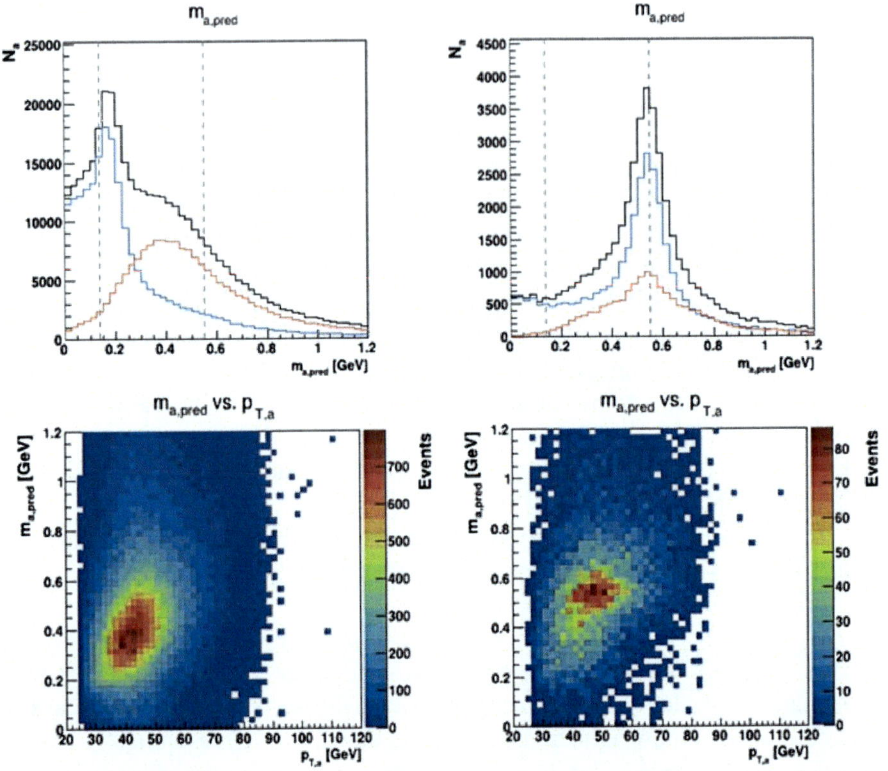

Fig. 7.5 Regressed m_Γ for jets matched to neutral meson decay in simulated γ + jet events. Upper: the total m_Γ spectrum (black) for jets matched to a generated $\pi^0 \to \gamma\gamma$ (left) or $\eta \to \gamma\gamma$ (right). The blue curve corresponds to the component of the m_Γ spectrum where a relatively isolated neutral meson decay is present, and the red curve, when otherwise. Compare to mass distributions in Fig. 7.4. Lower: the dependence of the non-isolated m_Γ component (red curve from upper plots) on the reconstructed p_{T}, for π^0-matched (left) and η-matched jets. Very loose event selection criteria are used, as described in Appendix A.2

at lower boosts (c.f. Fig. 7.4, top row). For the first of these, to understand how the collimation of additional hadrons affects the end-to-end versus the 3×3 algorithm, we select only samples reconstructed by the 3×3 algorithm within the mass window $400 < m_\Gamma < 600$ MeV. We then utilize only the 3×3 clusters of ECAL detector hits used by the 3×3 algorithm as input to the end-to-end m_Γ algorithm for evaluation. The results are shown in the left of Fig. 7.7, where we compare the selected m_Γ spectrum from the 3×3 method, with the height in the distribution subtracted (gray points, 3×3 clustering, no bg), to the corresponding spectrum from the end-to-end method, using either unmodified input detector hits (red circles, End-to-end) or only input hits associated with 3×3 clusters (blue cross, End-to-end, 3×3). We find that, when given only the 3×3 clusters as input, the impact of hadronization effects are substantially mitigated. Thus, when using the same reduced inputs, and when evaluated on an identical sample for which the 3×3 method is sensitive to the η

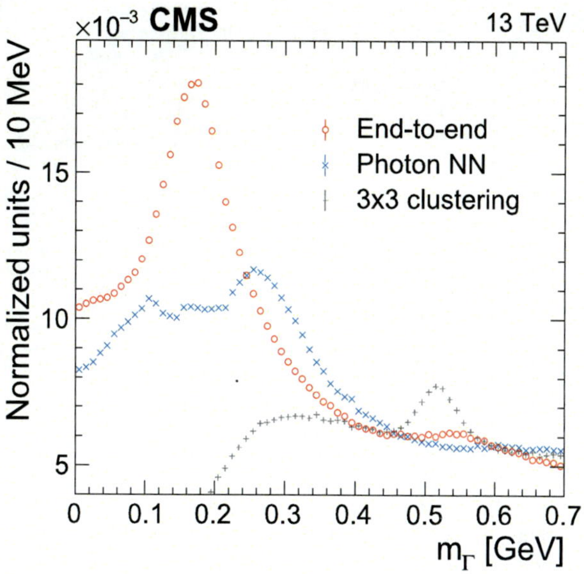

Fig. 7.6 Reconstructed mass m_Γ for end-to-end (red circles), photon NN (blue $\times$ symbol), and 3×3 (gray $+$ symbol) algorithms for hadronic jets from data enriched with $\pi^0 \rightarrow \gamma\gamma$ decays. All distributions are normalized to unity. See text for discussion. Credit: Ref. [4].

decay, the end-to-end method is also able to reconstruct a clear η resonance. For the second of these, as shown in the right of Fig. 7.7, for the η with a mass of 550 MeV, $p_T < 35$ GeV corresponds to $\gamma_L \lesssim 60$. This regime is at the edge of, or beyond, the trained phase space of the m_Γ regressor, where the reconstructed mass resolution may thus be less than optimal.

Whether the sensitivity of the end-to-end method to hadronization effects constitutes a positive or negative trait depends on the application. From the point of view of this analysis, since neutral meson decays are a background process, it is, in fact, beneficial for their decays in jets to be distinguished from those of isolated a $\rightarrow \gamma\gamma$. Indeed, the sensitivity of the end-to-end m_Γ regressor to hadronization effects puts it in a unique position to suppress neutral meson backgrounds in a way that would not be possible in an a $\rightarrow \gamma\gamma$ search performed with a lower energy collider where the mesons are well-isolated and hadronization effects are negligible [9]. Therefore, for the remainder of this analysis, the end-to-end method is employed using all detector hits as input, to maximize this effect. The impact of this effect on the modeling of the QCD background is described in Sect. 8.2.1. A more detailed analysis of this phenomenon, including the derivation of the results of Fig. 7.5, is presented in Appendix A.2. The optimization of the end-to-end mass regression technique for the particular phase space of $\eta \rightarrow \gamma\gamma$ decays in hadronic jets is left for future work.

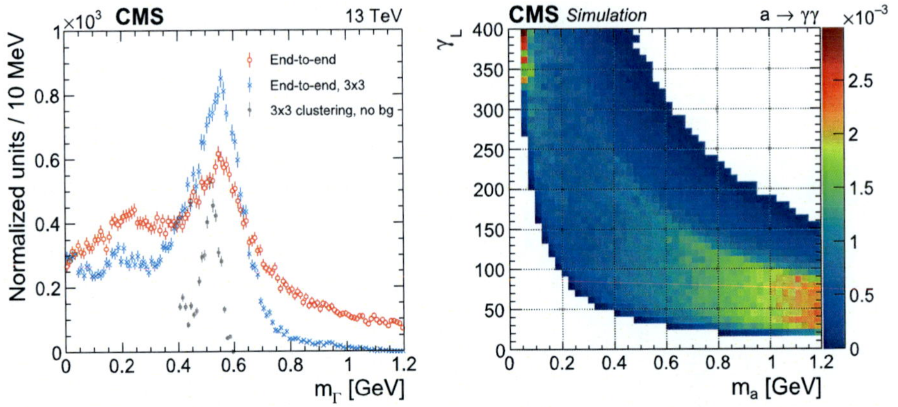

Fig. 7.7 Understanding the effect of jet collimation and low boost on the end-to-end ML regressed m_Γ.. Left: samples reconstructed by the 3×3 algorithm within the window $400 < m_\Gamma < 600$ MeV, comparing the mass spectrum obtained using the 3×3 algorithm, with the height in the distribution subtracted (gray points, 3×3 clustering, no bg), to that from the end-to-end method, using either unmodified input detector hits (red circles, End-to-end) or only input hits associated with 3×3 clusters (blue crosses, End-to-end, 3×3). Right: generated boosts γ_L vs m_{a} represented in training set. Boosts $\gamma_L > 400$ are not shown for clarity

7.5 Robustness of the Algorithm

Having validated the m_Γ regression strategy in general, in this section, we proceed to validate the regressor specifically for its application to the $\text{H} \to \text{aa} \to 4\gamma$ analysis. We measure the dependence of the m_Γ response to different years of data taking and detector conditions, as well as the dependence on simulation and data. Unless specified otherwise, we apply the same photon identification criteria used by the analysis, as described in Sect. 6.3. As stated above, neutral meson decays in jets are a background process for the $\text{a} \to \gamma\gamma$ search. Our photon identification criteria are thus designed to suppress QCD jet backgrounds, by exploiting the very same hadronic effects described above that distinguish their decays from those of isolated $\text{a} \to \gamma\gamma$.

7.5.1 $\pi^0 \to \gamma\gamma$ in Data

Despite the caveats associated with using neutral meson decays in jets, it is still useful to regress what can be of the $\pi^0 \to \gamma\gamma$ mass peak in data, as it is still a valuable tool for studying the mass dependence of the m_Γ regressor. In this subsection, we thus compare the dependence of the $\pi^0 \to \gamma\gamma$ mass peak in data on the different years of data taking. Note that, since the $\pi^0 \to \gamma\gamma$ is a background source for the this analysis, the photon criteria appropriate for $\text{a} \to \gamma\gamma$ selection (see Sect. 6) differs

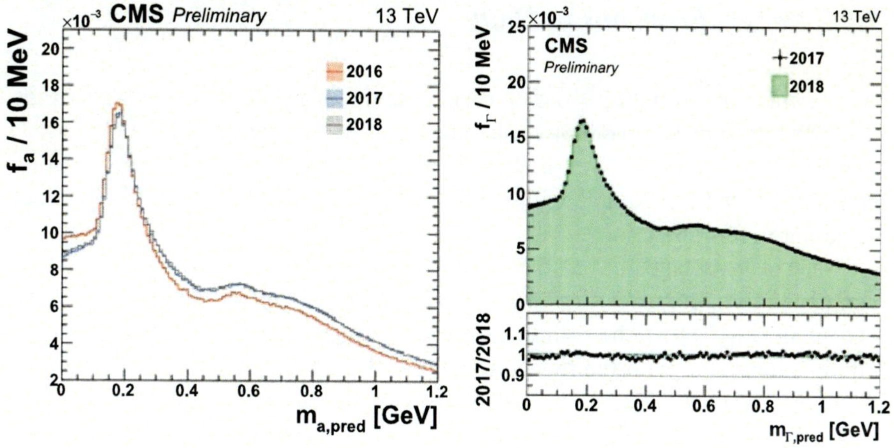

(a) 2018 vs. 2017 vs. 2016.

(b) 2018 vs. 2017.

Fig. 7.8 End-to-end reconstructed mass spectrum for subleading photon candidates passing event selection with reconstructed $p_{T,\Gamma} < 35$ and $55 < m_{\Gamma\Gamma} < 100$ GeV for the different years of data taking. 7.8: Overlay of spectrum from 2016, 2017, and 2018. 7.8: Overlay of spectrum from 2017 and 2018 for which years diphoton trigger was identical. Distributions are normalized to the total number of selected candidates

slightly from those used in the benchmark above, which are intended to enhance the $\pi^0 \to \gamma\gamma$ peak. In order to still mitigate the QCD effects described earlier, for this subsection alone, we require a few modifications to the kinematic requirements from Sect. 6. We instead use the lower-mass HLT diphoton trigger for each year of data taking[3], select events within $55 < m_{\Gamma\Gamma} < 100$ GeV, and reconstructed photons $p_{T,\Gamma} < 35$ GeV to maximize the component of isolated π^0s. As before, we regress only the subleading photon candidate to maximize contributions from jets in γ + jet events.

To compare the regressed m_Γ spectra between the different years of data taking, we first re-weight each year's data to have the same (p_T, η) distribution. Although this does not affect the $\pi^0 \to \gamma\gamma$ results significantly, the re-weighting is useful for the latter parts of this section and for consistency. The comparisons are shown in Fig. 7.8 for the three years.

We see that the $\pi^0 \to \gamma\gamma$ mass peak is consistently placed across the three years. In 2016, there is significantly higher photon content being selected, however, accounting for this, the position of the $\pi^0 \to \gamma\gamma$ is seen to be in good agreement with that of the other years. In Fig. 7.8, we focus specifically on the distributions for 2017 and 2018 to highlight the high level of agreement between them. This is especially notable given that the m_Γ regressor was trained exclusively on simulated isolated $a \to \gamma\gamma$ decays with 2017 detector conditions alone.

[3] 2016:`HLT_Diphoton30PV_18PV_R9Id_AND_IsoCaloId_AND_HE_R9Id_DoublePixelVeto_Mass55`. 2017, 2018: `HLT_Diphoton30PV_18PV_R9Id_AND_IsoCaloId_AND_HE_R9Id_PixelVeto_Mass55`.

7.5.2 e^+e^- Electrons in Data

As the comparison with 2016 data earlier showed, because of the effects of QCD, it is difficult to make quantitative statements about the stability of the regressed m_Γ unless a pure sample of well-isolated decays can be obtained. For this reason, we consider an alternate sample with which to measure the mass dependence of the m_Γ regressor in this analysis.

We take advantage of e^+e^- electrons which are produced in abundance with little contamination. Moreover, they are also produced well-isolated, similar to the way we expect the $a \to \gamma\gamma$s to be. While the electron is effectively massless at this regime, due its interaction with the magnetic field of the CMS solenoid, its shower profile is slightly smeared. This gives the electron shower a similar appearance to that of an extremely merged $a \to \gamma\gamma$ decay that exhibits completely unresolved shower maxima and subtle smearing to adjacent crystals. The electron m_Γ spectrum thus presents a peak at around $m_\Gamma \approx 100\,\text{MeV}$ that we can use to precisely quantify the mass scale. We emphasize that we intend to regress the m_Γ of *individual* electrons, not of the dielectron system, which would instead peak at the Z-boson mass.

In order to select e^+e^- electrons, we use a simplified version of the standard tag-and-probe method [10]. We require exactly two electron candidates in events passing the electron triggers prescribed for the tag-and-probe technique. The "tag" electron is identified by an electron candidate in the event passing the tight electron ID. A "probe" electron is selected if the other electron candidate in the event coincides with a photon candidate passing our identification requirements (minus the pixel veto) within $\Delta R < 0.04$, and forms a dielectron invariant mass of $60 < \mathrm{m_{ee}} < 120\,\text{GeV}$. If both photons pass the tight electron ID, both are selected as probes. Only the m_Γ of the selected probe electron is regressed. We re-weight the different years to have identical probe electron (p_T, η).

To estimate the difference in m_Γ response between a pair of years, we designate the earlier year the reference distribution and the latter year the test distribution.

We then parametrize the differences between the test and reference distributions in terms of a relative mass scale s_scale and a smearing difference s_smear. A scan is performed over different $(s_\text{scale}, s_\text{smear})$ hypothesis, in steps of $(0.002, 2\,\text{MeV})$. At each hypothesis, the transformation

$$m_\Gamma \to s_\text{scale} \times \mathcal{N}(m_\Gamma, s_\text{smear}), \tag{7.1}$$

is applied to each electron candidate $m_{\Gamma,\mathrm{i}}$ in the test sample, where $\mathcal{N}(\mu, \sigma)$ is a Gaussian function parametrized by mean μ and standard deviation σ. The best-fit mass scale $\hat{s}_\text{scale}$ and smearing $\hat{s}_\text{smear}$ between test and reference distributions is then defined as the hypothesis for which the chi-square (χ^2) test statistic between the mass distributions in the test and reference samples is at a minimum.

It should be acknowledged, however, that the regressor is fundamentally a non-linear function, whose mass response may vary in ways that are non-Gaussian. The

Table 7.2 Estimated difference in m_Γ scale and smearing between each pair of data taking years. Estimates are derived using e^+e^- electrons by comparing the χ^2 test statistic between the reference and test distribution under different scale & smearing hypotheses. Parameters are scanned in steps of ($\Delta s_{\text{scale}} = 0.002$, $\Delta s_{\text{smear}} = 2\,\text{MeV}$).

	2017 versus 2016	2018 versus 2017
$\hat{s}_{\text{scale}}$	0.950	0.996
$\hat{s}_{\text{smear}}$	18 MeV	8 MeV

above Gaussian parametrization, therefore, is to be understood as a best-attempt at approximating the variational modes of the regressor's response, in a simple way.

Table 7.2 shows the estimated scale and smearing differences in e^+e^- electrons using this procedure for 2017 versus 2016 and for 2018 versus 2017. To demonstrate how well these parametrizations account for the differences between the test and reference distributions, we determine the envelope between the best-fit and original test distributions ("ss"). After adding, in quadrature, the statistical uncertainties of the best-fit test distribution ("stat"), we plot the resulting envelope as a green band in the upper panels of Fig. 7.9, around the original test distribution (purple fill). The reference distribution is shown in black points. The parameters in Table 7.2 represent accurate uncertainty estimates of the mass scale and smearing between the test and reference distributions insofar as the green band encloses the reference distribution.

The lower panels of Fig. 7.9 plot the ratio of the reference over the original test distributions as black points, with error bars corresponding to the statistical uncertainties in the reference distribution. The ratio of the best-fit test distribution, plus statistical uncertainties, over the original test distribution, are shown as a green fill. In either upper or lower panels, we find the scale and smearing estimates to amply cover the differences between the test and reference distributions, justifying the choice of parametrization and estimation procedure.

We observe good stability in the m_Γ regressor's response in electrons across the years, measuring an agreement in the mass scale of 5% (< 1%), and smearing of 18 MeV (8 MeV), for 2017 versus 2016 (2018 vs. 2017), consistent with the qualitative findings from the $\pi^0 \rightarrow \gamma\gamma$ decays.

7.5.3 *Z e^+e^- Electrons in Data Versus Simulation*

As will be relevant for applying the simulation-derived signal model (see Sect. 8.1) on data, we require an estimate of the difference in the m_Γ regressor's response in data versus simulation. For the same reasons as in the previous section, we again use electrons, using the same procedure described earlier, to derive scale and smearing estimates that can be later used as systematics for the signal model. While the application of the scale and smearing systematics is treated formally in Sect. 8.3.7, we present here the derivation of these estimates. Note that the m_Γ scale and smearing

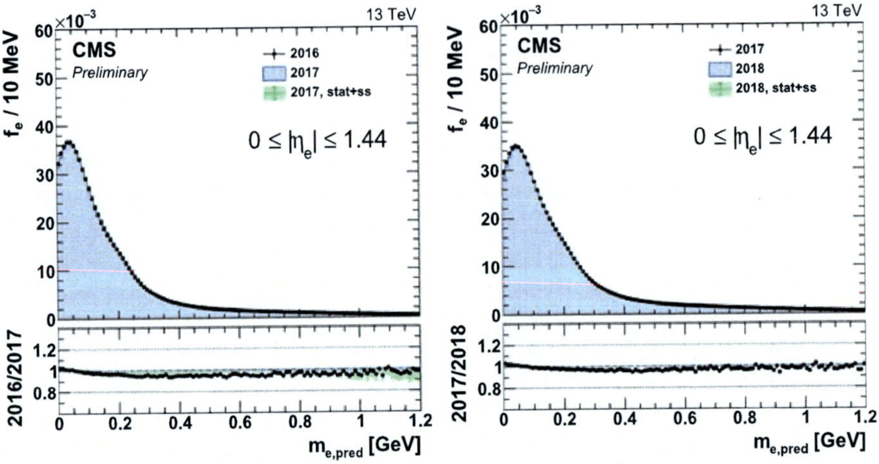

(a) 2017 vs 2016. (b) 2018 vs. 2017.

Fig. 7.9 End-to-end reconstructed mass spectrum for e^+e^- electrons passing photon identification for the different years of data taking. 7.9: 2017 vs 2016 data. 7.9: 2018 vs 2017. All distributions are normalized to unity. In the upper panels, the coverage of the best-fit scale and smearing estimates in the test distribution (later year, stat+ss), plus statistical uncertainties added in quadrature, are plotted as green bands around the original test distribution, shown in purple fill (later year). The reference distribution (earlier year) is shown as black points. In the lower panels, the ratio of test over the original reference distribution, is plotted as black points, with statistical uncertainties as error bars, and the ratio of best-fit to original test distribution, plus statistical uncertainties in the latter, is shown as a green fill

estimated here will not be used to correct the simulated m_Γ response, but rather, only used to estimate uncertainties for the m_Γ regressor (see Sect. 8.3.7).

Using the same electron tag-and-probe selection as described earlier, we treat the m_Γ spectrum derived from data as our reference distribution and that for simulation our test distribution. However, after re-weighting the probe electrons to have identical identical reconstructed (p_T, η), we split each sample by η range in order to obtain a finer description of the scale and smearing variations. A probe electron is categorized as either central ($|\eta| < 0.5$), middle ($0.5 \leq |\eta| < 1.0$), or forward ($1.0 \leq |\eta| < 1.44$) based on its reconstructed η. These boundaries are motivated by the differences in radiation lengths upon reaching the ECAL, due to underlying variations in the Tracker material budget. Additionally, the data versus simulation correction scale factors, described in Sect. 8.1, are first applied to the simulated sample. The resulting distributions are then used to determine the scale and smearing differences, for each year and η category, as described in the earlier procedure.

The estimated scale and smearing values for each year and η category are listed in Table 7.3. The coverage of each estimate is shown in Fig. 7.10, following a similar plotting strategy to that described earlier in Sect. 7.5.2: the data events are treated as the reference distribution (upper panel, black points) and the simulated events are treated as the test distribution (upper panel, purple fill), and the best-fit test distri-

Table 7.3 Estimated m_Γ regressor scale and smearing differences between data and simulation, derived separately for different years and η ranges. Parameters are scanned in steps of ($\Delta s_{\text{scale}} = 0.002$, $\Delta s_{\text{smear}} = 2\,\text{MeV}$)

η		2016	2017	2018
$\|\eta\| < 0.5$	$\hat{s}_{\text{scale}}$	1.004	1.046	1.012
	$\hat{s}_{\text{smear}}$	6 MeV	–	2 MeV
$0.5 \leq \|\eta\| < 1.0$	$\hat{s}_{\text{scale}}$	0.978	1.032	1.018
	$\hat{s}_{\text{smear}}$	–	–	–
$1.0 \leq \|\eta\| < 1.44$	$\hat{s}_{\text{scale}}$	1.002	1.056	1.048
	$\hat{s}_{\text{smear}}$	10 MeV	–	–

bution, plus statistical uncertainties, as an envelope (green fill) around the original test distribution. The last of these is labelled `MC, stat+syst` in anticipation of its later use as a systematic for the signal model. Similarly, in the lower panels of Fig. 7.10, the ratio between data and simulation (MC), plus statistical uncertainties in the former, is plotted as black points, and the ratio between the best-fit and original simulation, plus statistical uncertainties in the former, is shown as a green fill.

For most plots, we see mass scale agreement within a few percent, and little to no smearing. This suggest that the regressor, non-linear as it is, is able to absorb most of the discrepancy through just a scale difference. An important consequence of this is that the full mass resolution of the simulated signal models will be manifest in the analysis, with no reduction in signal significance. The positions of the mass peaks are also seen to be consistent between data and simulation, with any discrepancies often well-accounted for by the scale and smearing differences. For a number of plots, disagreements in the high-side tail slightly beyond what is covered by the scale and smearing estimates, can be seen. These are likely due to differences in the modeling of low energy detector hits, which are more apparent in AOD rechits, and which tend to drive the shape of the mass tails. In this analysis, these segments of the m_Γ spectrum will be dominated by statistical uncertainties (see Sect. 8.1) and are thus not expected to have a significant impact on the signal modeling.

7.5.4 *a → γγ in Simulated H → aa → 4γ*

Lastly, we look at the dependence of the regressed m_Γ in $\text{a} \rightarrow \gamma\gamma$ decays on year-to-year detector conditions, using simulated $\text{H} \rightarrow \text{aa} \rightarrow 4\gamma$ signal samples. For each year, the m_Γ spectrum is decomposed by leading and subleading $p_{\text{T},\Gamma}$ as defined by the analysis event selection (see Sect. 6.4). These are shown in Fig. 7.11 for different generated m_{a}.

Again, we see excellent agreement in the m_Γ response across the years for different generated masses and $p_{\text{T},\Gamma}$ ranges. As mentioned earlier, this is especially notable given the m_Γ regressor was trained exclusively on 2017 simulation.

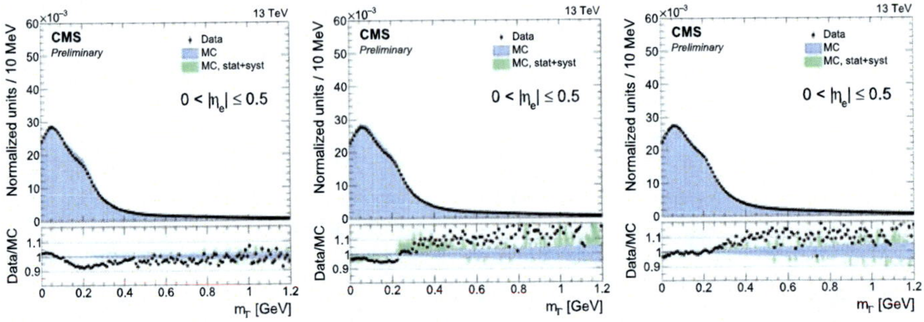

(a) Central, $|\eta| < 0.5$.

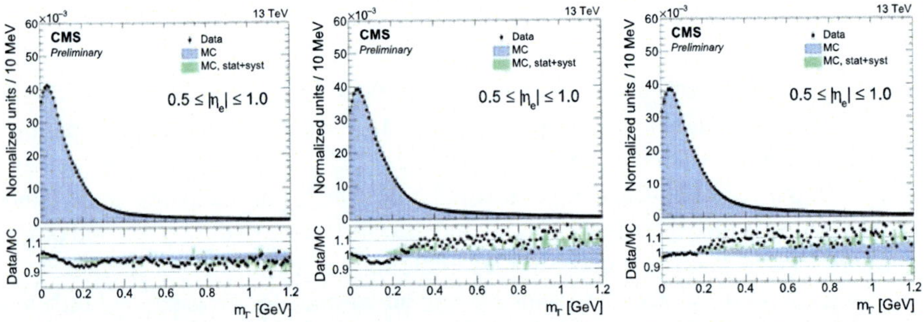

(b) Middle, $0.5 \leq |\eta| < 1.0$.

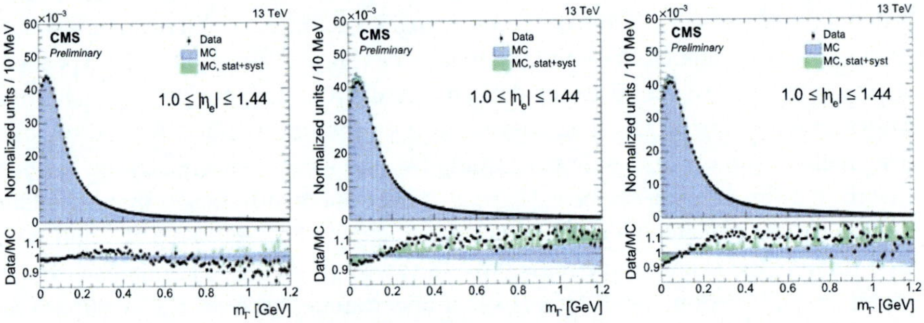

(c) Forward, $1.0 \leq |\eta| < 1.44$.

Fig. 7.10 End-to-end reconstructed mass for e^+e^- electrons in data versus simulation for the years 2016 (left column), 2017 (middle column), and 2018 (right column) by η range: central (7.10a), middle (7.10b), forward (7.10c). All distributions are normalized to unity. In the upper panels of each plot, the coverage of the best-fit scale and smearing estimates in simulated events (MC, stat+syst), plus statistical uncertainties added in quadrature, are plotted as green bands around the original simulated sample, shown in purple fill (MC). The data events are shown as black points (Data), with statistical uncertainties as error bars. In the lower panels of each plot, the ratio of data over simulation is plotted as black points, with statistical uncertainties in the former as error bars, and the ratio of best-fit to original simulated distribution, plus statistical uncertainties in the former, is shown as a green fill.

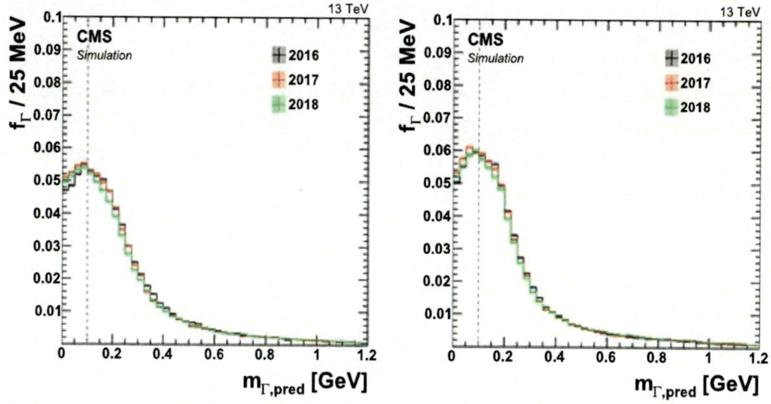

(a) $m_a = 100\,\text{MeV}$.

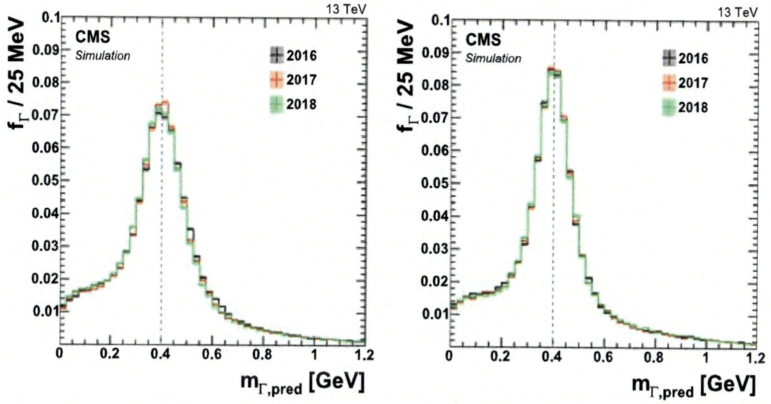

(b) $m_a = 400\,\text{MeV}$.

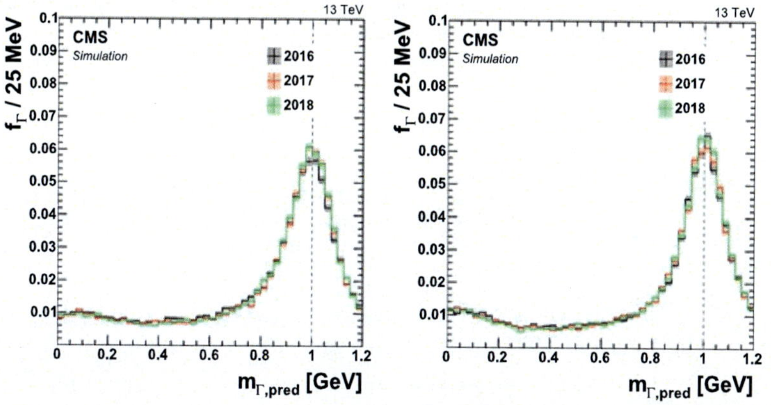

(c) $m_a = 1\,\text{GeV}$.

Fig. 7.11 End-to-end reconstructed mass spectrum for simulated $\text{H} \rightarrow \text{aa} \rightarrow 4\gamma$ events passing event selection for different years of simulated data-taking conditions. The spectra shown for generated masses of $m_a = 100\,\text{MeV}$ (7.11a), $m_a = 400\,\text{MeV}$ (7.11b), and $m_a = 1\,\text{GeV}$ (7.11c). The mass spectra for the leading (subleading) $p_{T,\Gamma}$ candidate is shown on the left (right) column. All distributions are normalized to unity

In this section, we have shown how the end-to-end ML technique can be used to achieve breakthrough sensitivity in $a \rightarrow \gamma\gamma$ mass reconstruction over a wide range of previously inaccessible boosts. We showed that we were able to probe the most challenging yet attractive mass regime for which the $a \rightarrow \gamma\gamma$ decays is completely unresolved. Additionally, this was achieved with a robustness to different kinematic regimes and detector conditions.

References

1. Andrews M, Paulini M, Gleyzer S, Poczos B (2019) Exploring end-to-end deep learning applications for event classification at cms. EPJ Web Conf 214:06031, 01 2019
2. Andrews M, Paulini M, Gleyzer S, Poczos B (2020) End-to-end physics event classification with cms open data: Applying image-based deep learning to detector data for the direct classification of collision events at the lhc. Comput Softw Big Sci 4(1):6
3. Andrews M, Alison J, An S, Burkle B, Gleyzer S, Narain M, Paulini M, Poczos B, Usai E (2020) End-to-end jet classification of quarks and gluons with the cms open data. Nucl Instrum Methods Phys Res Sect Accelerators, Spectrometers, Detectors Associated Equip. 164304
4. Reconstruction of decays to merged photons using end-to-end deep learning with domain continuation in the CMS detector (2022)
5. Kingma DP, Ba J (2014) Adam: a method for stochastic optimization
6. Cowan G, Cranmer K, Gross E, Vitells O (2011) Asymptotic formulae for likelihood-based tests of new physics. Eur Phys J C 71
7. Khachatryan V et al (2015) Performance of photon reconstruction and identification with the CMS detector in proton-proton collisions at $\sqrt{s} = 8$TeV. JINST 10:P08010
8. Khachatryan V et al (2014) Observation of the diphoton decay of the Higgs boson and measurement of its properties. Eur Phys J C 74(10):3076
9. Adhikari S et al (2022) Search for photoproduction of axionlike particles at GlueX. Phys Rev D 105:052007
10. Chatrchyan S, Khachatryan V, Sirunyan AM, Tumasyan A, Adam W, Bergauer T, Dragicevic M, Ero J, Fabjan C et al (2011) Measurement of the inclusive w and z production cross sections in pp collisions at $\sqrt{s} = 7$tev with the cms experiment. J High Energy Phys 2011, Oct 2011

Chapter 8
Analysis

In this chapter, we describe the main components of the physics analysis associated with the $H \rightarrow aa \rightarrow 4\gamma$ signal search, namely the signal and background models. These are the primary pieces that will be used to test for the presence of a signal in the data in the following chapter. As described in Chap. 4, both these models are expressed in the 2D-m_Γ observable, to enable a direct and physically compelling discrimination of the $H \rightarrow aa \rightarrow 4\gamma$ signal, if it exists. The signal model, which encapsulates the differential response expected in the data if the $H \rightarrow aa \rightarrow 4\gamma$ decay were realized, is described in Sect. 8.1. The background model, which describes the expected SM-only response in data, absent the $H \rightarrow aa \rightarrow 4\gamma$ signal, is described in Sect. 8.2. Lastly, the so-called systematics, or the uncertainties associated with the construction of these models, are detailed in Sect. 8.3.

8.1 Signal Model

As described in Sect. 4.3, the signal model is expressed in terms of the 2D-m_Γ distribution or "template" of simulated $H \rightarrow aa \rightarrow 4\gamma$ events. In order for the signal model to be properly compared to data in the final signal region m_H-SR $\cap$ m_a-SR, we require both an estimate of the signal model's intrinsic 2D-m_Γ shape at a given mass hypothesis, as well as its total yield or normalization with respect to a data sample of a given integrated luminosity.

To derive the shape of the signal model, a 2D-m_Γ template is filled using simulated $H \rightarrow aa \rightarrow 4\gamma$ signal events passing the selection criteria described in Sect. 6. Note that the Higgs boson is produced inclusively of all SM production modes (see Sect. 5.2). The 2D-m_Γ templates are generated for each mass hypothesis m_a to be evaluated against the background model. Dedicated samples are generated for

M. Andrews, *Search for Exotic Higgs Boson Decays to Merged Diphotons*, Springer Theses, https://doi.org/10.1007/978-3-031-25091-0_8

the masses $m_{\mathrm{a}} = 0.1, 0.2, 0.4, 0.6, 0.8, 1.0, 1.2\,\mathrm{GeV}$. An interpolation procedure, described below, is applied to cover the remaining mass points in the range $m_{\mathrm{a}} = [0.1, 1.2]\,\mathrm{GeV}$ in 0.1 GeV steps.

Having derived the shape of the 2D-m_Γ distribution for a particular mass hypothesis, to derive its yield $N(m_{\mathrm{a}}|m_H\text{-SR} \cap m_{\mathrm{a}}\text{-SR})$ in the m_H-SR $\cap$ m_{a}-SR, for a given integrated luminosity of data $\int \mathcal{L}\,dt$, the following conversion is used,

$$N(m_{\mathrm{a}}|m_H\text{-SR} \cap m_{\mathrm{a}}\text{-SR}) = \int \mathcal{L}\,dt \times \sigma(\mathrm{H}) \times \mathcal{B}(\mathrm{H} \rightarrow \mathrm{aa} \rightarrow 4\gamma) \times \varepsilon(m_{\mathrm{a}}|m_H\text{-SR} \cap m_{\mathrm{a}}\text{-SR}), \tag{8.1}$$

where $\sigma(\mathrm{H})$ is the Higgs boson production cross section, $\mathcal{B}(\mathrm{H} \rightarrow \mathrm{aa} \rightarrow 4\gamma)$ is the branching fraction to $\mathrm{H} \rightarrow \mathrm{aa} \rightarrow 4\gamma$, and $\varepsilon(m_{\mathrm{a}}|m_H\text{-SR} \cap m_{\mathrm{a}}\text{-SR})$ is the efficiency with which $\mathrm{H} \rightarrow \mathrm{aa} \rightarrow 4\gamma$ events of a given m_{a} hypothesise are selected into the m_H-SR $\cap$ m_{a}-SR. For computational reasons, we assume a hypothetical signal cross section times branching fraction of $\sigma(\mathrm{H}) \times \mathcal{B}(\mathrm{H} \rightarrow \mathrm{aa} \rightarrow 4\gamma) = 1\,\mathrm{pb}$ to normalize the 2D-m_Γ templates. This is equivalent to a $\mathcal{B}(\mathrm{H} \rightarrow \mathrm{aa} \rightarrow 4\gamma) = 2 \times 10^{-2}$ for the total inclusive SM Higgs production cross section of $\sigma(\mathrm{H}) = 51.04\,\mathrm{pb}$ [1].

Unfortunately, physically motivated estimates for the full $BR(\mathrm{H} \rightarrow \mathrm{aa} \rightarrow 4\gamma)$ are difficult to come by. Due to the non-perturbative QCD regime involved in $m_{\mathrm{a}} \lesssim 1\,\mathrm{GeV}$ (see Sect. 3.3.1), theoretical estimates for $\mathcal{B}(\mathrm{a} \rightarrow \gamma\gamma)$ break down at the masses considered in this analysis [2]. Predictions for $\mathcal{B}(\mathrm{a} \rightarrow \gamma\gamma)$ in this regime are thus mostly qualitative, motivated by the remaining available decay modes of the particle a to $3\pi^0$s, dimuons, dielectrons, or diphotons. However, if the a is only allowed to couple to the Higgs and heavy, vector-like, uncolored states at the renormalizable level, it may only decay to diphotons [2]. If we assume optimistically that $\mathcal{B}(\mathrm{a} \rightarrow \gamma\gamma) \approx 1$, then a total branching fraction of $\mathcal{B}(\mathrm{H} \rightarrow \mathrm{aa} \rightarrow 4\gamma) = 2 \times 10^{-2}$ would fall within conservative theoretical estimates for the $\mathrm{H} \rightarrow \mathrm{aa}$ piece of the branching fraction [2]. Realistically, however, for highly-merged $\mathrm{a} \rightarrow \gamma\gamma$ topologies ($m_{\mathrm{a}} \lesssim 400\,\mathrm{MeV}$), model assumptions giving signal branching fractions greater than those of SM $\mathrm{H} \rightarrow \gamma\gamma$, $\mathcal{B}(\mathrm{H} \rightarrow \mathrm{aa} \rightarrow 4\gamma) \gtrsim \mathcal{B}(\mathrm{H} \rightarrow \gamma\gamma) \sim 2 \times 10^{-3}$, are likely to already be excluded by SM $\mathrm{H} \rightarrow \gamma\gamma$ coupling measurements [3]. At higher m_{a}, on the other hand, the looser photon MVA requirement in our photon identification requirements, relative to that of the SM $\mathrm{H} \rightarrow \gamma\gamma$ analysis, leaves more room for discovery.

The approximate size of the signal model for different m_{a} hypotheses is shown in Figs. 8.1, 8.2 and 8.3, including the corrections described in the following subsections. The 2D-m_Γ distribution in the m_{a}-SR region is shown in the uppermost plot. The unrolled 2D-m_Γ distribution is in the center plot, and the 1D-m_Γ projections for $m_{\Gamma,1}$ ($m_{\Gamma,2}$) are shown in the lower-left (-right) plots.

In the upper panels of the center and lower plots, the black points ($\mathrm{H} \rightarrow \mathrm{aa} \rightarrow 4\gamma$) outlining the gray region denote the signal model, with error bars corresponding to statistical uncertainties. The systematic uncertainties (Sg, syst) associated with the signal model are shown as a green band around the black points. For reference, the background model is shown as a blue line, with its statistical uncertainties indicated by the blue band. In the lower panels of the same plots, the ratio of the statistical (gray band) and systematic (green band) uncertainties in the signal model, over its nominal value is displayed.

As the mass peak falls off toward the tails of the distribution, systematic uncertainties begin to dominate the signal model, as expected. A description of the systematic uncertainties associated with the signal model is deferred to Sect. 8.3. As described in Sect. 4.3, the final signal region for which the signal model will be required corresponds to the region of overlap between the signal regions in $m_{\Gamma\Gamma}$ and 2D-m_{Γ}, m_H-SR $\cap$ m_{a}-SR.

These plots are also indicative of the gain in signal sensitivity achieved by using the 2D-m_{Γ} spectra over the individual 1D-m_{Γ} spectra: the background contribution within the vicinity of the signal peak is seen to be slightly lower for the 2-D case.

In addition to the $\mathrm{H} \to \mathrm{aa} \to 4\gamma$ signal model thus derived, a number of corrections must be implemented to account for the fact these have been derived from simulation but are being compared to data. Due to residual mismodeling of the detector response in simulation relative to data, a number of effects arise that impact the yield and shape of the 2D-m_{Γ} templates derived from simulation. These include differences in selection efficiency under the photon identification criteria, the HLT diphoton trigger, and the pileup modeling. In the following subsections, we discuss how the signal model is modified to account for these effects.

8.1.1 Photon Identification Scale Factors

Due to residual mismodeling of the detector response in simulation, the efficiency with which photon candidates pass our identification criteria (see Chap. 6) in simulation may slightly differ from that in data. To correct for this, we apply photon identification *scale factors* (SFs), representing the relative selection efficiency in data versus simulation, as a function of (p_{T}, η). These SFs can then be used to re-weight each signal event used to fill the 2D-m_{Γ} signal model based on where each of its photon candidates lie in (p_{T}, η).

To estimate these SFs, we use $\mathrm{Z} \to \mathrm{e}^+\mathrm{e}^-$ events collected in both data and simulation using a single electron HLT trigger. We then use the tag-and-probe method [4] to obtain a pure sample of unbiased (probe) electrons in both data and simulation. From these, we can calculate the number electrons passing and failing the photon identification criteria (minus the pixel veto requirement) as a function of (p_{T}, η) in data, and separately, in simulation. The scale factor is then the ratio of the selection efficiencies in data versus simulation at a given (p_{T}, η) bin. The SFs are shown in the lower panel of Fig. 8.4 for the different years of data-taking, as a function of p_{T} and η. The total weight for the signal event is given by the product of the SFs of each photon candidate.

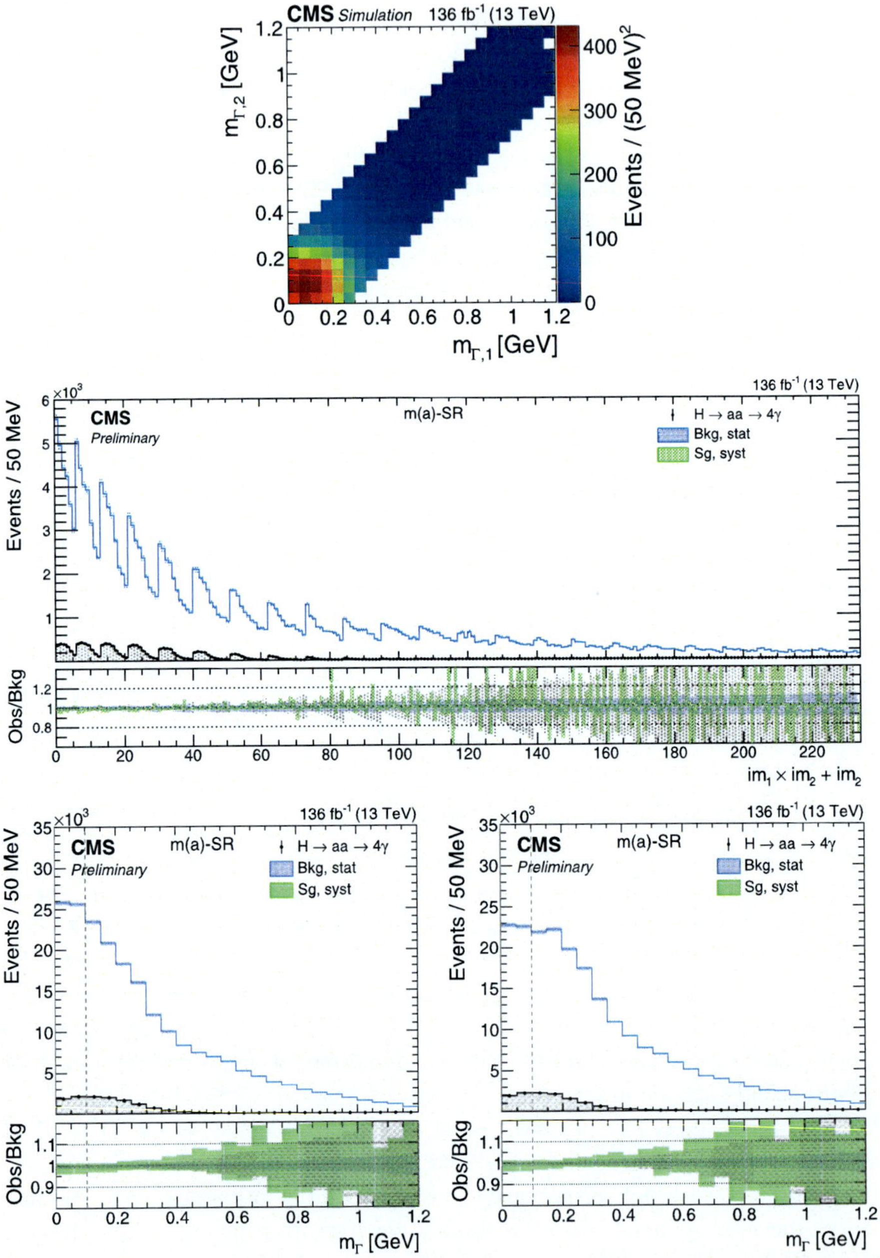

Fig. 8.1 Simulation-derived H $\rightarrow$ aa $\rightarrow 4\gamma$ signal model for the $m_{\mathrm{a}} = 100\,\mathrm{MeV}$ mass hypothesis, normalized to $\sigma(\mathrm{H}) \times \mathcal{B}(\mathrm{H} \rightarrow \mathrm{aa} \rightarrow 4\gamma) = 1\,\mathrm{pb}$, in the m_{a}-SR. Upper: 2D-m_{Γ} distribution. Center: unrolled 2D-m_{Γ} distribution. Lower: projected 1D-m_{Γ} distributions for $m_{\Gamma,1}$ (left) and $m_{\Gamma,2}$ (right). In the upper panels of the center and lower plots, the black points (H $\rightarrow$ aa $\rightarrow 4\gamma$) outlining the gray region denote the signal model, with error bars corresponding to statistical uncertainties. The associated systematic uncertainties (Sg, syst) are shown as a green band. The background model is shown as a blue line, with its statistical uncertainties indicated by the blue band. In the lower panels of the same plots, the ratio of the statistical (gray band) and systematic (green band) uncertainties in the signal model, over the nominal values are displayed

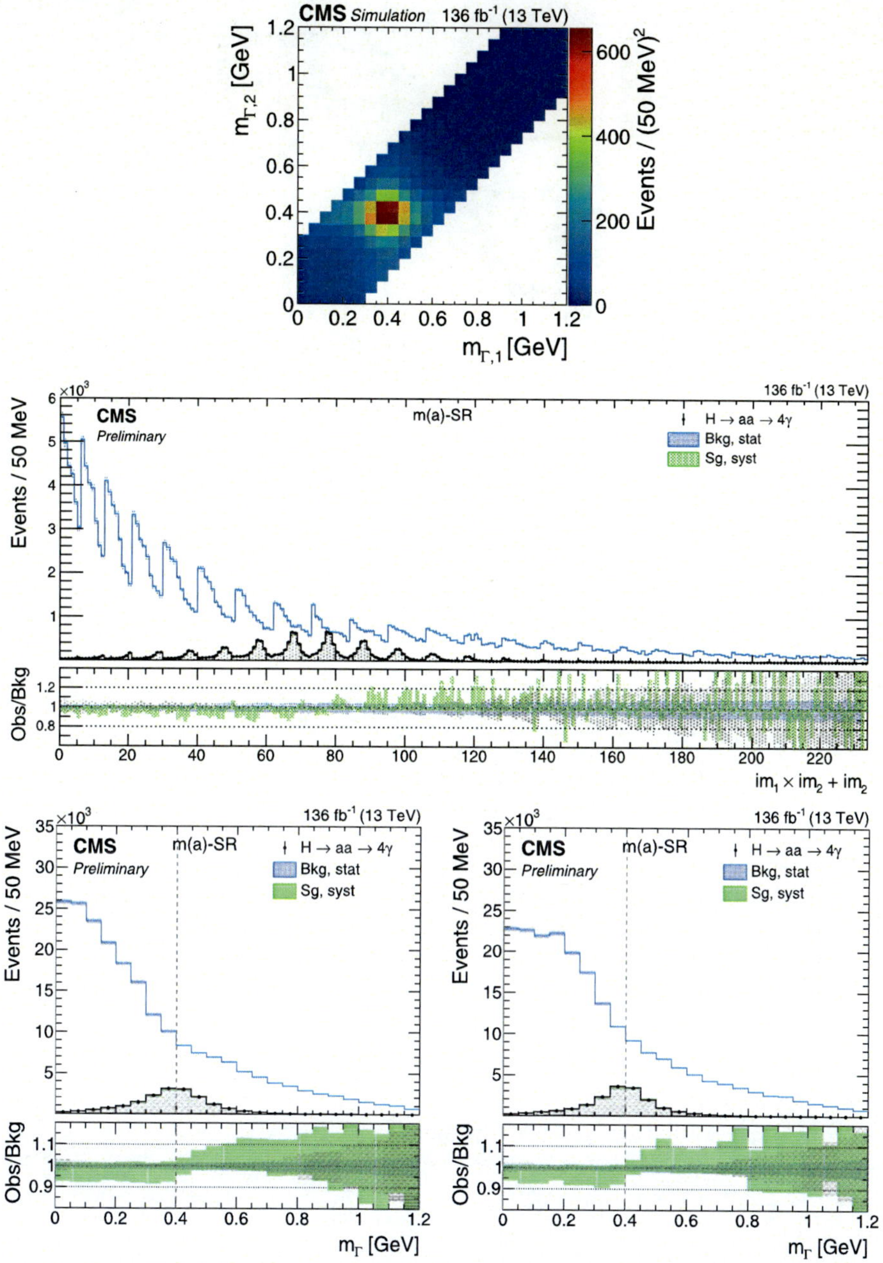

Fig. 8.2 Simulation-derived $\mathrm{H} \rightarrow \mathrm{aa} \rightarrow 4\gamma$ signal model for the $m_{\mathrm{a}} = 400\,\mathrm{MeV}$ mass hypothesis, normalized to $\sigma(\mathrm{H}) \times \mathcal{B}(\mathrm{H} \rightarrow \mathrm{aa} \rightarrow 4\gamma) = 1\,\mathrm{pb}$, in the m_{a}-SR. Upper: 2D-m_{Γ} distribution. Center: unrolled 2D-m_{Γ} distribution. Lower: projected 1D-m_{Γ} distributions for $m_{\Gamma,1}$ (left) and $m_{\Gamma,2}$ (right). In the upper panels of the center and lower plots, the black points ($\mathrm{H} \rightarrow \mathrm{aa} \rightarrow 4\gamma$) outlining the gray region denote the signal model, with error bars corresponding to statistical uncertainties. The associated systematic uncertainties (Sg, syst) are shown as a green band. The background model is shown as a blue line, with its statistical uncertainties indicated by the blue band. In the lower panels of the same plots, the ratio of the statistical (gray band) and systematic (green band) uncertainties in the signal model, over the nominal values are displayed

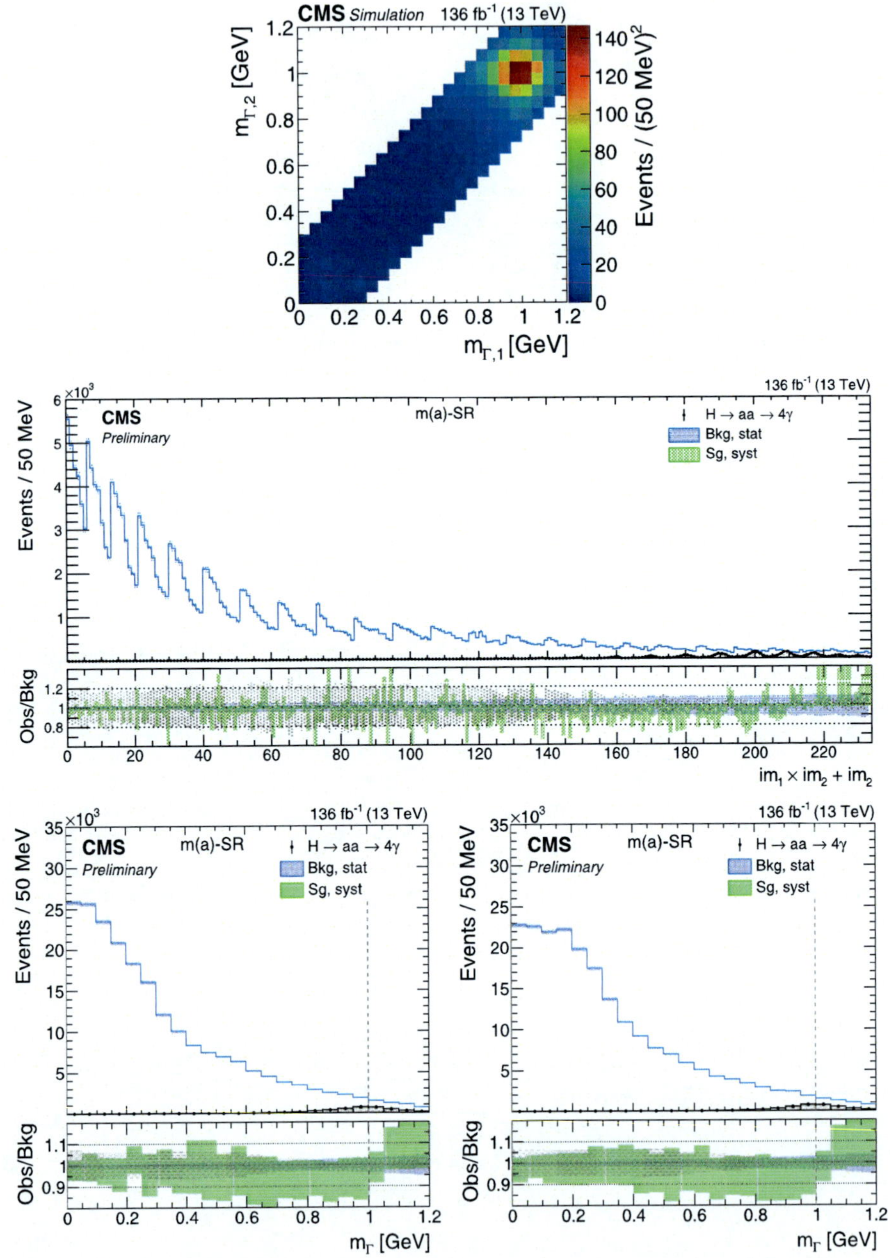

Fig. 8.3 Simulation-derived $\mathrm{H} \to \mathrm{aa} \to 4\gamma$ signal model for the $m_{\mathrm{a}} = 1\,\mathrm{GeV}$ mass hypothesis, normalized to $\sigma(\mathrm{H}) \times \mathcal{B}(\mathrm{H} \to \mathrm{aa} \to 4\gamma) = 1\,\mathrm{pb}$, in the m_{a}-SR. Upper: 2D-m_{Γ} distribution. Center: unrolled 2D-m_{Γ} distribution. Lower: projected 1D-m_{Γ} distributions for $m_{\Gamma,1}$ (left) and $m_{\Gamma,2}$ (right). In the upper panels of the center and lower plots, the black points ($\mathrm{H} \to \mathrm{aa} \to 4\gamma$) outlining the gray region denote the signal model, with error bars corresponding to statistical uncertainties. The associated systematic uncertainties (Sg, syst) are shown as a green band. The background model is shown as a blue line, with its statistical uncertainties indicated by the blue band. In the lower panels of the same plots, the ratio of the statistical (gray band) and systematic (green band) uncertainties in the signal model, over the nominal values are displayed

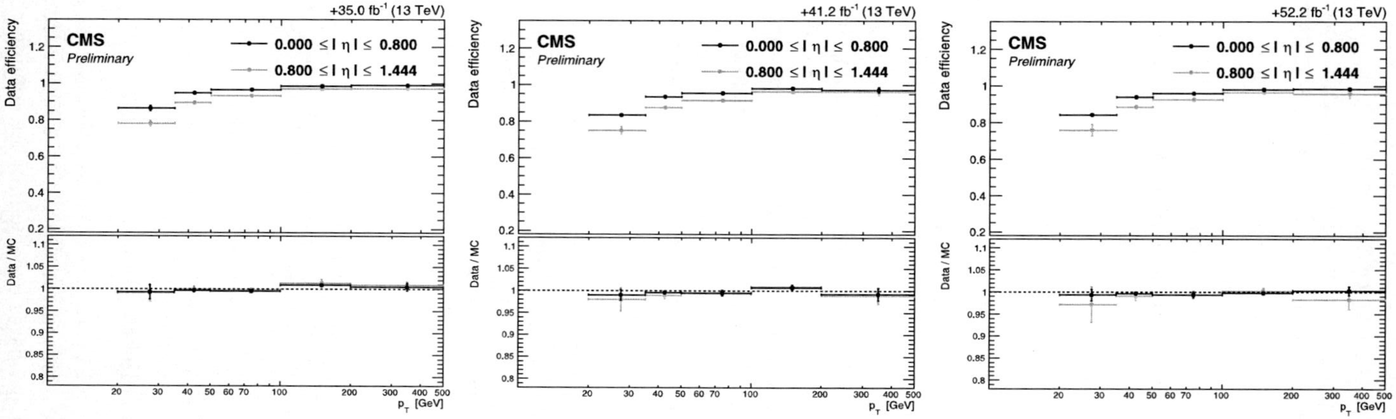

Fig. 8.4 Photon ID efficiencies and scale factors as a function of p_T and η, as derived for 2016 (left), 2017 (center), and 2018 (right) data-taking

The number of passing and failing electron probes are determined using analytic fits, which carry a fit uncertainty, and for which we introduce a systematic uncertainty in the shape of the signal model (see Sect. 8.3).

While this derivation necessarily relies on electrons, because of the bending of electrons in the magnetic field of the CMS solenoid, electron showers are slightly smeared, and thus exhibit a range of shower shapes that approximately capture those of $a \rightarrow \gamma\gamma$ decays at various masses.

The photon identification SFs used in this analysis were derived separately for this analysis, using the the criteria described in Sect. 6.3.

8.1.2 HLT Trigger Scale Factors

A separate but related impact of the residual mismodeling of the detector response in simulation versus data is the difference in efficiency with which $H \rightarrow aa \rightarrow 4\gamma$ event pass the HLT diphoton trigger in data versus simulation. This is further complicated by the changes in the underlying L1 paths used to seed the HLT diphoton trigger, which may change throughout each year of data-taking in response to changes in the LHC beam luminosity. This makes emulating the response of the HLT trigger particularly challenging, especially near the photon p_T thresholds or at "turn-on".

To account for these effects, rather than try to emulate the HLT diphoton trigger directly in simulation, we instead measure the HLT trigger selection efficiency in data using an alternate, unbiased sample and apply these as scale factors to the simulated signal events. As before, an unbiased sample of electrons is used, obtained using the tag-and-probe method. In order to decouple the effect due to the trigger from that due to the photon identification, the trigger efficiencies are measured in data after requiring the probe electrons to pass photon identification criteria (minus the pixel veto). The trigger efficiencies are measured as a function of p_T, η, and the shower shape variable R_9. These are subsequently applied to each photon candidate, for each event in the simulated sample, in addition to the photon identification SFs.

The number of passing and failing electron probes are also determined using analytic fits, which carry a fit uncertainty, and for which we introduce another systematic uncertainty in the shape of the signal model (see Sect. 8.3).

For simplicity, the trigger SFs used in this analysis are those derived for the SM $H \rightarrow \gamma\gamma$ analysis using the $H \rightarrow \gamma\gamma$ photon preselection, which are similar to the photon identification criteria used in this analysis.

8.1.3 Pileup Re-weighting

The PU distribution or scenario under which a simulated sample is generated is typically a projection of what is expected for the coming year of data taking. Over the course of data collection, however, the realities of data taking typically mean the actual PU distribution observed for the year may differ slightly from projections. As the amount of PU in an event can potentially affect the amount of electromagnetic activity in the ECAL (among other subsytems), this, in turn, can impact the efficiency for an event to pass our selection criteria. To correct for this, the simulated signal events used to derive the signal templates are first re-weighted in order that their sample-wide PU distribution agrees with that observed for the relevant year. The weights are derived using normalized histograms of the PU distribution in data versus simulation for each year of data-taking. The ratio between data and simulation at each bin of PU gives the PU weight for that bin. In practice, this correction is sub-dominant, with most of the PU event weights within $<1\%$ of unity.

To cross-check the PU re-weighting procedure, we compare the distribution in the number of reconstructed primary vertices per event in data versus the PU re-weighted simulated samples. This quantity is closely correlated with the number of PU interactions and is used as a complementary check. Comparison distributions are shown in Fig. 8.5 for different signal model mass points. Since the simulated samples are generated and PU re-weighted separately for each year, these comparisons are presented separately for 2016 (Fig. 8.5a), 2017 (Fig. 8.5b), and 2018 (Fig. 8.5c) data-taking. We observe good agreement across all samples. While some deviations are seen, particularly for the tails of the distributions, these are to be expected given that the PU weights are derived for finite PU bin widths.

8.1.4 Sample Interpolation

To obtain a finer scan in the m_a hypotheses when testing for the presence of a signal (see Sect. 9.1), we apply an interpolation procedure to derive intermediate-mass signal models given the available dedicated samples described earlier. To obtain the signal model for an interpolated mass hypothesis m_B between two generated mass hypotheses m_A and m_C, with $m_A < m_B < m_C$, the following procedure is followed:

- For each regressed mass $m_{A,i}$ in the sample $\{m_A\}$ generated with mass hypothesis m_A, scale the mass value by

$$m_{A,i} \rightarrow \left(\frac{m_B}{m_A}\right) m_{A,i}. \tag{8.2}$$

  The resulting sample with scaled masses is then $\{m_{A\rightarrow B}\}$. The same procedure is applied for sample $\{m_C\}$ to yield the scaled sample $\{m_{C\rightarrow B}\}$.
- Generate the raw 2D-m_Γ distribution for the combined scaled samples $\{m_{A\rightarrow B} \cup m_{C\rightarrow B}\}$
- Normalize the 2D-m_Γ as in Eq. 8.1 except using the selection efficiency ε_{sg} defined over the combined scaled sample $\{m_{C\rightarrow B}\}$.

A sample signal model for an interpolated mass point at $m_a = 500$ MeV is shown in Fig. 8.6. In addition, to provide an estimate of the worst-case scenario achieved under the mass interpolation scheme, in Fig. 8.7, we compare the m_Γ distribution from

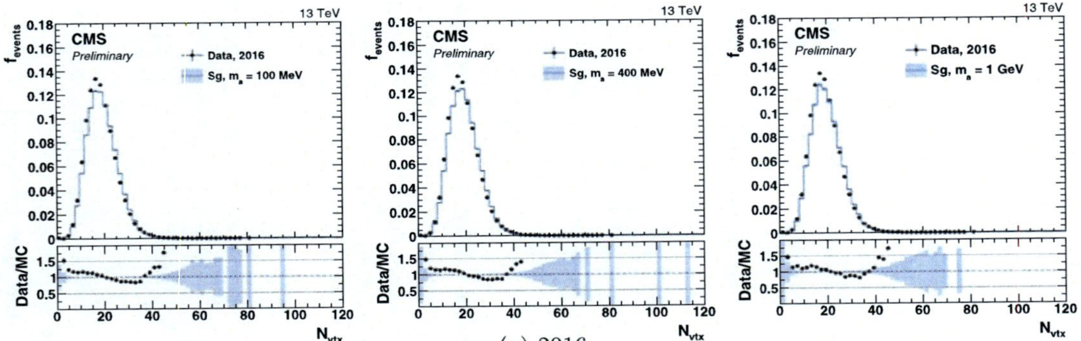

(a) 2016.

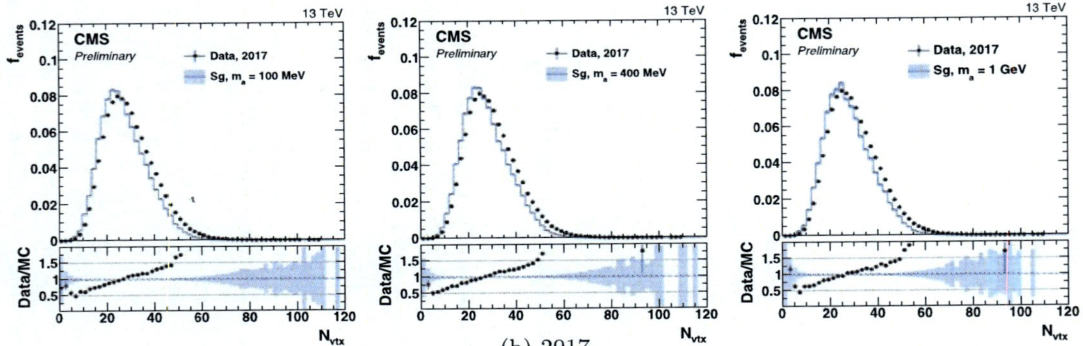

(b) 2017.

Fig. 8.5 Distributions in number of reconstructed primary vertices per event in data (black points) versus simulation (blue line) for signal model mass points $m_{\mathrm{a}} = 100$ MeV (left column), 400 MeV (middle column), and 1 GeV (right column). Shown separately for data-taking years 2016 (**a**), 2017 (**c**), and 2018 (**c**). All distributions are normalized to unity

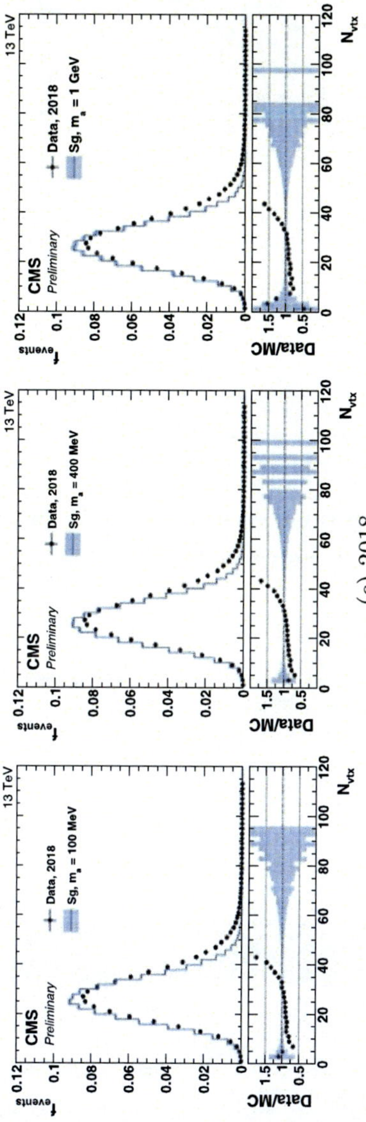

(c) 2018.

Fig. 8.5 (continued)

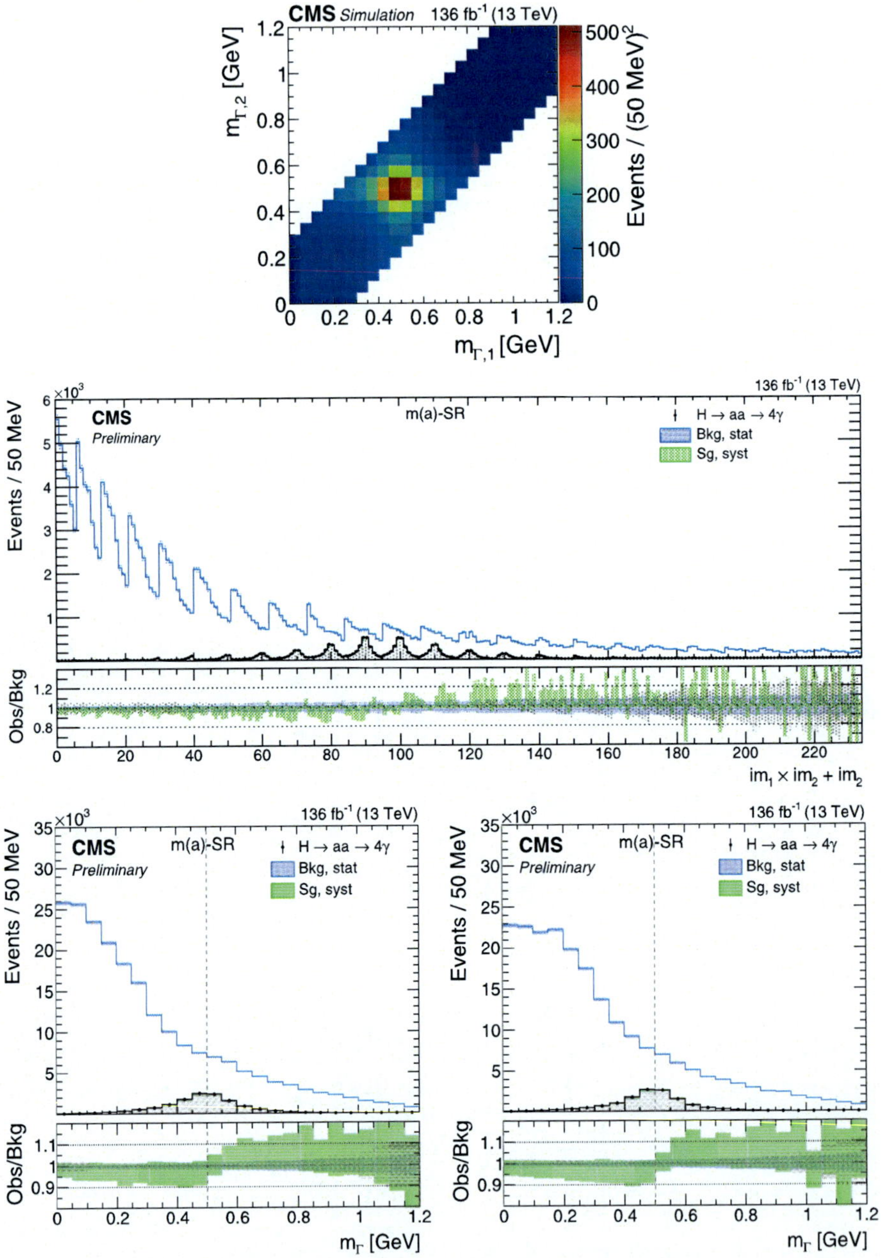

Fig. 8.6 Interpolated $H \to aa \to 4\gamma$ signal model for the $m_a = 500$ MeV mass hypothesis, normalized to $\sigma(H) \times \mathcal{B}(H \to aa \to 4\gamma) = 1$ pb, in the m_a-SR. Upper: 2D-m_Γ distribution. Center: unrolled 2D-m_Γ distribution. Lower: projected 1D-m_Γ distributions for $m_{\Gamma,1}$ (left) and $m_{\Gamma,2}$ (right). In the upper panels of the center and lower plots, the black points ($H \to aa \to 4\gamma$) outlining the gray region denote the signal model, with error bars corresponding to statistical uncertainties. The associated systematic uncertainties (Sg, syst) are shown as a green band. The background model is shown as a blue line, with its statistical uncertainties indicated by the blue band. In the lower panels of the same plots, the ratio of the statistical (gray band) and systematic (green band) uncertainties in the signal model, over the nominal values are displayed.

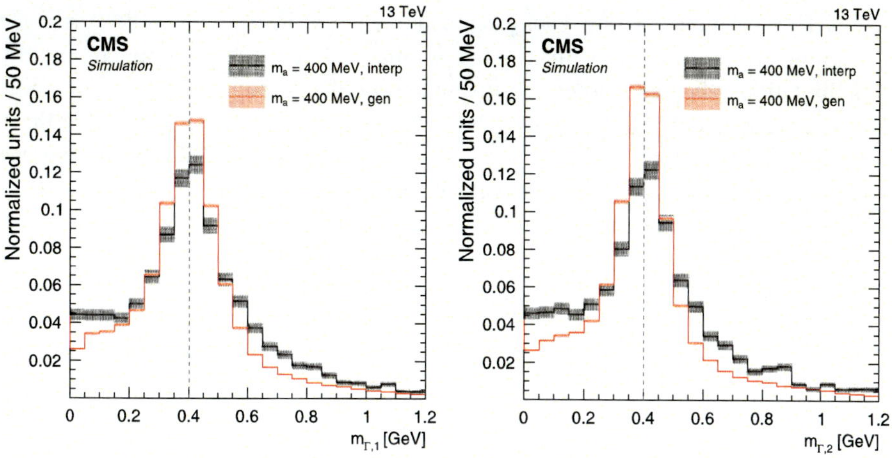

Fig. 8.7 The regressed m_Γ spectrum for H → aa → 4γ signal events generated at $m_a = 400$ MeV (gen, red), versus that interpolated (interp, gray) from $m_a = 200$ and 600 MeV. Left: leading photon candidate $m_{\Gamma,1}$. Right: subleading photon candidate $m_{\Gamma,2}$. All distributions normalized to unity

an interpolated mass point, with that of a generated one, for $m_a = 400$ MeV. This comparison necessarily involves interpolating between more distant mass points, in this case, between $m_a = 200$ and 600 MeV. As such, the comparison is not intended to be representative of the actual mass distributions achieved in the interpolated mass points used in the analysis, which involve half the distance.

8.2 Background Model

The leading SM background sources for the analysis, in order of decreasing cross-section, are QCD multijet and γ + jet with EM-enriched jets, prompt diphoton production, and the H → $\gamma\gamma$ decay, as shown earlier in Fig. 3.10. Except for the H → $\gamma\gamma$ process, which has a resonance at $m_{\Gamma\Gamma} = 125$ GeV, all the other background sources are produced non-resonantly with a smoothly falling spectrum in $m_{\Gamma\Gamma}$. While these do not take into account event selection efficiencies, even so, QCD sources in the form of partons hadronizing to $\pi^0 \rightarrow \gamma\gamma$ decays or radiating energetic photons (final state radiation, FSR) overwhelmingly dominate the data sample selected by our signal selection criteria in Chap. 6. This section describes the procedure for estimating the contribution of these processes to the 2D-m_Γ spectrum in the region over which the H → aa → 4γ signal search will be performed.

For all but the H → $\gamma\gamma$ contribution, a data-driven approach is used to construct a total non-resonant SM background 2D-m_Γ template. Data sidebands in both $m_{\Gamma\Gamma}$ and 2D-m_Γ (see Sect. 4.3) are exploited to derive the shape and yield of this background template. For the H → $\gamma\gamma$ contribution, which falls directly over the m_H-SR, simulated data is used to derive the shape of the background template, while theoretical predictions are used to determine its yield. The final signal region for which the data-driven plus H → $\gamma\gamma$ background model must be constructed is the region of overlap between both signal regions, m_H-SR ∩ m_a-SR.

Before deriving the full background model, it is instructive to first describe a simplified procedure for estimating the dominant, data-driven component, highlight its limitations, then use these to motivate the full background estimation procedure.

The simplified data-driven background estimation is as follows: we assume that the intrinsic 2D-m_Γ shape in the m_H-SR signal region $p(m_a|m_H\text{-SR})$ can be expressed as a linear sum of the 2D-m_Γ shapes in the lower and upper m_H-SB sidebands, $p(m_a|m_H\text{-SB}_{\text{low}})$ and $p(m_a|m_H\text{-SB}_{\text{high}})$, respectively,

$$p(m_\Gamma|m_H\text{-SR}) = p(m_\Gamma|m_H\text{-SB}), \tag{8.3}$$

where

$$p(m_\Gamma|m_H\text{-SB}) \equiv f_{\text{SB-low}}\, p(m_\Gamma|m_H\text{-SB}_{\text{low}}) + f_{\text{SB-high}}\, p(m_\Gamma|m_H\text{-SB}_{\text{high}}), \tag{8.4}$$

$$f_{\text{SB-low}} + f_{\text{SB-high}} = 1. \tag{8.5}$$

The fractions $f_{\text{SB-low}}$ ($f_{\text{SB-high}}$) are the relative contributions of the m_H-SB$_{\text{low}}$ (m_H-SB$_{\text{high}}$) shape to the total m_H-SB shape, $p(m_a|m_H\text{-SB})$. Since the m_H-SB is signal-depleted by construction, we can derive the relevant shapes from these sidebands, and combine them under some suitable choice of the fractions, to determine the total shape, $p(m_\Gamma|m_H\text{-SB})$.

For the normalization, we assume that the ratio in the number of events along the 2D-m_Γ diagonal, from that observed in the m_H-SRs to that derived in the m_H-SB, $N(m_H\text{-SR} \cap m_a\text{-SR})/N(m_H\text{-SB} \cap m_a\text{-SR})$, is the same as for the off-diagonal, $N(m_H\text{-SR} \cap m_a\text{-SB})\,/N(m_H\text{-SB} \cap m_a\text{-SB})$,

$$\frac{N(m_H\text{-SR} \cap m_a\text{-SR})}{N(m_H\text{-SB} \cap m_a\text{-SR})} = \frac{N(m_H\text{-SR} \cap m_a\text{-SB})}{N(m_H\text{-SB} \cap m_a\text{-SB})} \tag{8.6}$$

From the off-diagonal region, which is signal-depleted by construction, we can derive the ratio from the right hand side (RHS) of the above equation, in addition to the m_a-SR yield in the m_H-SB. The yield in the final signal region, m_H-SR $\cap$ m_a-SR, is then determined to be

$$N(m_H\text{-SR} \cap m_a\text{-SR}) = \frac{N(m_H\text{-SR} \cap m_a\text{-SB})}{N(m_H\text{-SB} \cap m_a\text{-SB})} N(m_H\text{-SB} \cap m_a\text{-SR}). \tag{8.7}$$

We thus define the data-driven background model in the final signal region as the shape derived from Eq. 8.3, scaled by the yield from Eq. 8.7,

$$\begin{aligned} \text{bkg}(\text{data}, m_H\text{-SR} \cap m_a\text{-SR}) &\equiv N(m_H\text{-SR} \cap m_a\text{-SR})\, p(m_\Gamma|m_H\text{-SR}) \\ &= N(m_H\text{-SB} \cap m_a\text{-SR})\, p(m_\Gamma|m_H\text{-SB} \cap m_a\text{-SR}) \\ &\quad \times \frac{N(m_H\text{-SR} \cap m_a\text{-SB})}{N(m_H\text{-SB} \cap m_a\text{-SB})} \end{aligned} \tag{8.8}$$

This simple background model, however, has two main limitations. The first is the missing $\mathrm{H} \to \gamma\gamma$ contribution. Since the $\mathrm{H} \to \gamma\gamma$ decay is a resonant background, it meets none of these assumptions—by construction, it is absent from the m_H sidebands. We must, therefore, introduce an additional $\mathrm{H} \to \gamma\gamma$ background template, using simulated events. The second limitation is due to the p_T-dependent m_Γ spectrum of QCD jets. Because of this, the background naively derived from the $m_{\Gamma\Gamma}$ sidebands will not accurately model the m_Γ shape of jets in the m_H-SR. To motivate our solution, it is helpful to first understand how this issue arises. Thus, in the following subsection, we take a closer look at the substructure and topology of hadronic jets that drive this effect. We return to a description of the complete background estimation procedure in the subsection thereafter.

8.2.1 QCD Jet Substructure

The 2D-m_Γ background shape is primarily driven by the underlying particle species and their p_T spectrum. The m_Γ spectrum of a photon-dominated sample exhibits a smoothly falling spectrum that is mostly p_T-independent [5]. However, as discussed earlier in Sect. 7.4.2, a mostly QCD jet sample, on the other hand, exhibits a more complicated shape due to the dependence of the physical jet topology on p_T. At low-p_T, jets passing the photon identification criteria are more likely to contain only a single isolated $\pi^0 \to \gamma\gamma$ which are regressed as a (p_T-independent) $m_\mathrm{a} \approx 135\,\mathrm{MeV}$ peak. However, as the jet grows more energetic at higher p_T, more energy becomes available to produce additional, potentially boosted, hadrons. In order to pass photon identification, selected jets are thus more likely to contain two or more sets of very collimated $\pi^0 \to \gamma\gamma$s ($\Delta R \lesssim 1$ ECAL crystal in order to pass shower shape cuts), which are regressed as a broad spectrum around $m_\mathrm{a} \approx 500\,\mathrm{MeV}$ that is p_T-dependent (more π^0s means more effective mass), as shown in Fig. 8.8. A more detailed analysis of this phenomenon, including the derivation of the results of Fig. 8.8, is presented in Appendix A.1.

In general, the 2D-m_Γ shape interpolated from the m_H-SBs will not have the correct underlying p_T and so the shape of the QCD jet component will not be accurately modeled. This can be mitigated in two ways while still maintaining a data-driven background. The first is to minimize the QCD contribution so that the effect is small to begin with. The second is to re-weight the underlying p_T distribution of the m_H-SB events so that they match what is found in the m_H-SR. We apply both of these. The photon identification criteria, as described in Chap. 6, already provide strong suppression of the most critical, p_T-dependent component of the QCD background. We find no further reduction to be possible without sacrificing signal efficiency, short of developing a dedicated $\mathrm{a} \to \gamma\gamma$ tagger. For the p_T re-weighting, we rely on the fact that the m_Γ spectrum of the non-critical, p_T-independent components of the background will be unchanged under the p_T re-weighting, or at least will not cause the background modeling to worsen if the p_T modeling is improved (c.f. Fig. 8.8, left column, for jets and Fig. 7.4, bottom row, for photons). We then assume that the

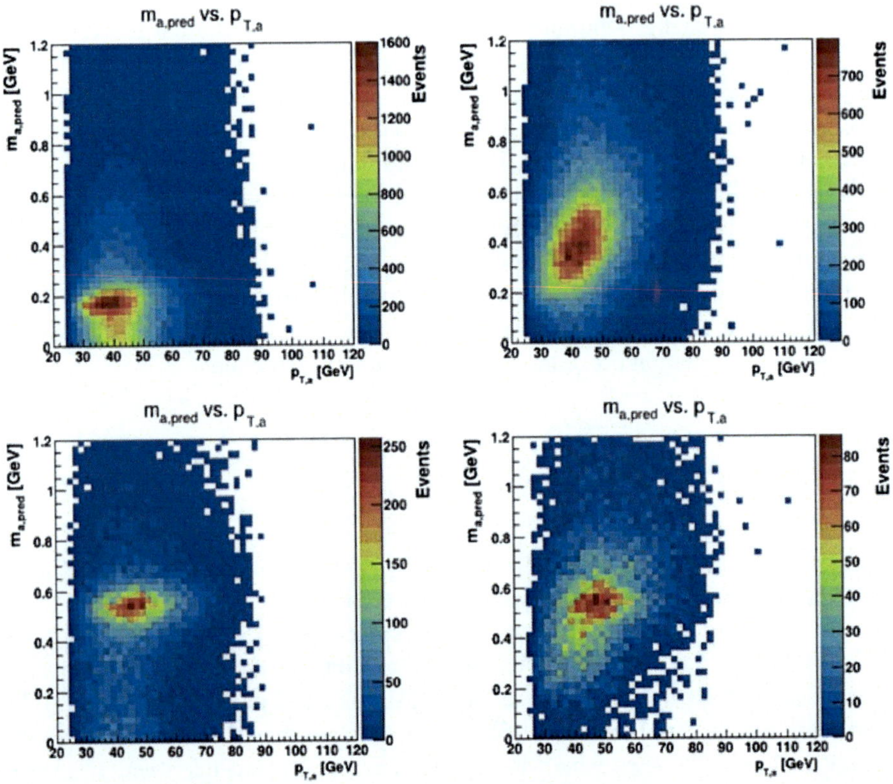

Fig. 8.8 Dependence of the regressed m_Γ on reconstructed p_T for electromagnetically enriched jets in simulated γ + jet events. Left: reconstructed m_Γ vs p_T for jets matched to a generated $\pi^0 \to \gamma\gamma$ (upper) or $\eta \to \gamma\gamma$ (lower), for neutral meson decays which are relatively isolated in the jet. Right: the corresponding distributions when the neutral meson decays are not isolated. Very loose event selection criteria are used, as described in Appendix A.1

total p_T in each data region which we use for the p_T re-weighting will be sufficient to correct the QCD-only p_T distribution that is the source of the mismodeling. Within the limited simulated QCD data available, we have verified that this assumption is indeed valid. We include a systematic to cover our statistical confidence in this statement (see Sect. 8.3.1).

8.2.2 Background Estimation

Having discussed the limitations of the simplified background estimation procedure, and the appropriate modifications to correct its shortcomings, we now describe in more concrete terms the full background estimation procedure. These steps are outlined below, in the order in which they are performed. Appropriate visualizations are provided to guide the discussion.

p_T re-weighting Due to the sensitivity of the regressed m_Γ spectrum to variations in the QCD jet topology over the p_T spectrum (c.f. Fig. 8.8), the best background modeling is achieved when the m_H-SB events used to derive the 2D-m_Γ templates are first re-weighted to have the same $p_{T,2}$ versus $p_{T,1}$ (2D-p_T) distribution as events in the m_H-SR. The re-weighting is performed from the total, normalized 2D-p_T probability distribution in the combined m_H-SBs, $p(p_T|m_H\text{-SB})$, to the total, unblinded, normalized 2D-p_T in the m_H-SR, $p(p_T|m_H\text{-SB})$. While it is primarily the QCD component of the 2D-p_T that necessitates the p_T re-weighting, we find it a suitable approximation to use the total 2D-p_T , which is valid insofar as the QCD-only and total 2D-p_T distributions are similar. We verify this to be the case, within uncertainties, in Sect. 8.3.1, where we introduce a systematic reflecting our confidence in this statement. We note that even the components of the background which we identify as p_T-independent (single photons, isolated π^0), exhibit a weak p_T dependence (c.f. Fig. 8.8, left column, for jets and Fig. 7.4, bottom row, for photons), smearing more with increasing p_T. They thus only benefit from the total 2D-p_T re-weighting.

If we denote the normalized 2D-p_T probability distribution of each m_H-SB i as $p(p_T|m_H\text{-SB}_i)$, then, as illustrated in Fig. 8.9, the combined probability distribution is given in general form as

$$p(p_T|m_H\text{-SB}) = f_{\text{SB-low}}\, p_T(p_T|m_H\text{-SB}_{\text{low}}) + f_{\text{SB-high}}\, p_T(p_T|m_H\text{-SB}_{\text{high}}) \tag{8.9}$$

for some choice of the relative fractions $f_{\text{SB-low (high)}}$. In order to maximize the statistical power of the events from each sideband, we choose

$$f_{\text{SB-low}} = \frac{N_{\text{SB-low}}}{N_{\text{SB-low}} + N_{\text{SB-high}}} \tag{8.10}$$

where $N_{\text{SB-low (high)}}$ is the total event yield in the $m_H\text{-SB}_{\text{low (high)}}$ sideband. This is equivalent to adding the raw 2D-p_T distributions of each m_H-SB and normalizing it to the yield in the m_H-SR afterwards. The systematic we define for this p_T re-weighting procedure is expressed in terms of an uncertainty in the choice of the parameter $f_{\text{SB-low}}$ (see Sect. 8.3).

After carrying out this procedure, the p_T re-weighted 2D-m_Γ shape templates from either $m_H\text{-SB}_{\text{low (high)}}$ are added to obtain the combined m_H-SB shape template, which we denote as $p_{\text{rwgt}}(m_\Gamma|m_H\text{-SB})$. This is the p_T re-weighted counterpart of Eq. 8.4. Note that the fraction that each m_H-SB contributes to the combined 2D-m_Γ shape template is fixed by the fraction $f_{\text{SB-low}}$ used in the p_T re-weighting, as given in Eq. 8.10.

2D-m_Γ normalization To obtain the data-driven background model, all that remains is to determine the normalization. The arguments used to derive Eq. 8.8 remain valid in this case, from which we determine the p_T re-weighted data-driven background model in the final signal region to be

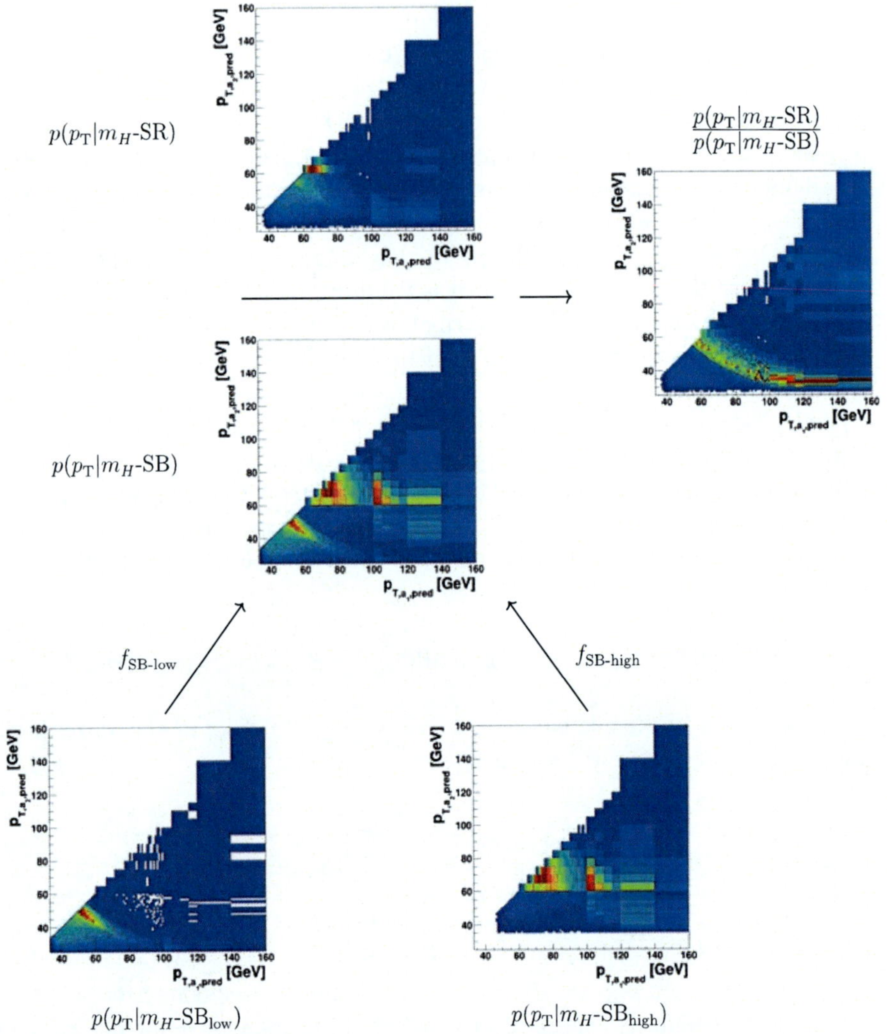

Fig. 8.9 Illustration of the p_{T} re-weighting procedure: the p_{T} shapes in the m_H-SB$_{\mathrm{low}}$ (lower left) and m_H-SB$_{\mathrm{high}}$ (lower right) are added together with fractions $f_{\text{SB-low}}$ and $f_{\text{SB-high}}$, respectively, to obtain a combined m_H-SB shape (denominator). This is divided from the p_{T} shape in the m_H-SR (numerator), to determine the p_{T} re-weighting factor as a function of an event's 2-D p_{T} (top right). Binning is chosen to ensure reasonable statistics

$$\mathrm{bkg}_{\mathrm{rwgt}}(\mathrm{data}, m_\Gamma | m_H\text{-SR}) = N(m_H\text{-SR} \cap m_{\mathrm{a}}\text{-SR})\, p_{\mathrm{rwgt}}(m_\Gamma | m_H\text{-SB} \cap m_{\mathrm{a}}\text{-SR}), \tag{8.11}$$

where $N(m_H\text{-SR} \cap m_{\mathrm{a}}\text{-SR})$ is as given in Eq. 8.7 and $p_{\mathrm{rwgt}}(m_\Gamma | m_H\text{-SB} \cap m_{\mathrm{a}}\text{-SR})$ is the shape obtained in the m_H-SB $\cap$ m_{a}-SR subset of the above $p_{\mathrm{rwgt}}(m_\Gamma | m_H\text{-SB})$. The above procedures of template combination and normalization are illustrated in Fig. 8.10.

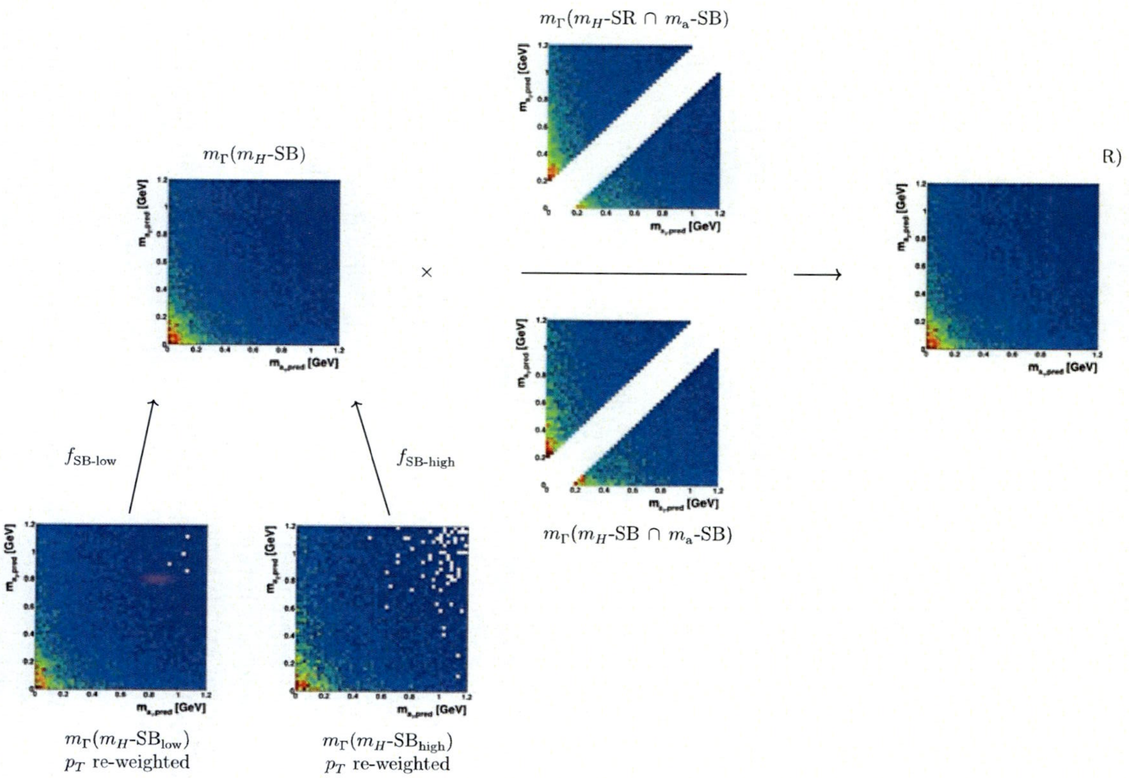

Fig. 8.10 Derivation and normalization of the data-driven 2D-m_Γ background model: the re-weighted m_Γ distributions in the m_H-SB$_{\text{low}}$ and m_H-SB$_{\text{high}}$ (lower left plots) are added together with fractions $f_{\text{SB-low}}$ and $f_{\text{SB-high}}$, respectively, to obtain the combined m_H-SB template (middle left). The yield in the combined template is then multiplied by the ratio of the yields in the m_H-SR $\cap$ m_a-SB (numerator) over the m_H-SB $\cap$ m_a-SB (denominator) to give the p_T re-weighted, data-driven background model in the m_H-SR $\cap$ m_a-SR (upper right)

H → γγ contribution To account for the contribution of the $\mathrm{H} \to \gamma\gamma$ process in the final signal region, an additional 2D-m_Γ template, bkg($\mathrm{H} \to \gamma\gamma$, $m_\Gamma|m_H$-SR $\cap$ m_{a}-SR), must be added to the above data-driven background model. Since the $\mathrm{H} \to \gamma\gamma$ contribution is contained in the m_H-SR, it cannot be completely derived from data. Instead, the $\mathrm{H} \to \gamma\gamma$ 2D-m_Γ template is derived in two parts. The first part, the intrinsic 2D-m_Γ shape $p(\mathrm{H} \to \gamma\gamma,\ m_\Gamma|m_H\text{-SR} \cap m_{\mathrm{a}}\text{-SR})$, is derived from simulation, while the second part, the normalization $N(\mathrm{H} \to \gamma\gamma,\ m_\Gamma|m_H\text{-SR} \cap m_{\mathrm{a}}\text{-SR})$, is derived from theory.

We use simulated $\mathrm{H} \to \gamma\gamma$ events passing our event selection to determine the shape $p(\mathrm{H} \to \gamma\gamma,\ m_\Gamma|m_H\text{-SR})$ over the full 2D-m_Γ plane. For simplicity, only events from the gluon fusion Higgs production mode are used, though this is the dominant contribution (>80%). We do not expect the 2D-m_Γ shape from the other Higgs production modes to differ significantly as the m_Γ response for photons is relatively consistent [5].

To determine the normalization $N(\mathrm{H} \to \gamma\gamma,\ m_{\mathrm{a}}|m_H\text{-SR} \cap m_{\mathrm{a}}\text{-SR})$ corresponding to a given integrated luminosity of data, we first parametrize the normalization as a fraction $f_{\mathrm{H}\to\gamma\gamma}$ of the total data yield in the final signal region,

$$N(\mathrm{H} \to \gamma\gamma,\ m_{\mathrm{a}}|m_H\text{-SR} \cap m_{\mathrm{a}}\text{-SR}) = f_{\mathrm{H}\to\gamma\gamma} N(m_\Gamma|m_H\text{-SR} \cap m_{\mathrm{a}}\text{-SR}). \quad (8.12)$$

To derive $f_{\mathrm{H}\to\gamma\gamma}$, having obtained the shape $p(\mathrm{H} \to \gamma\gamma,\ m_\Gamma|m_H\text{-SR})$ from simulation, we modify the assumption in Eq. 8.3 to read

$$p_{\mathrm{data,rwgt+H}\to\gamma\gamma}(m_\Gamma|m_H\text{-SR}) = f_{\mathrm{SB}}\, p_{\mathrm{rwgt}}(m_\Gamma|m_H\text{-SB}) + f_{\mathrm{H}\to\gamma\gamma}\, p(\mathrm{H} \to \gamma\gamma,\ m_\Gamma|m_H\text{-SR}), \quad (8.13)$$

with

$$f_{\mathrm{SB}} + f_{\mathrm{H}\to\gamma\gamma} = 1. \quad (8.14)$$

The fraction f_{SB} is the relative contribution of the data-driven 2D-m_Γ shape. Importantly, we have assumed in Eq. 8.13 that the equality holds identically over the m_{a}-SR and m_{a}-SB regions of the 2D-m_Γ plane, which is justified insofar as the 2D-m_Γ shape varies proportionally between the two regions. We can thus derive $f_{\mathrm{H}\to\gamma\gamma}$ from the signal-depleted m_{a}-SB counterpart of Eq. 8.12 instead. Using the counterpart of Eq. 8.1 for $\mathrm{H} \to \gamma\gamma$ events, and substituting this into the m_{a}-SB counterpart of Eq. 8.12, we obtain

$$f_{\mathrm{H}\to\gamma\gamma} = \frac{1}{N(m_\Gamma|m_H\text{-SR} \cap m_{\mathrm{a}}\text{-SB})} \times \int \mathcal{L}\,\mathrm{d}t \times \sigma(\mathrm{H}) \times \mathcal{B}(\mathrm{H} \to \gamma\gamma) \times \varepsilon(\mathrm{H} \to \gamma\gamma,\ m_{\mathrm{a}}|m_H\text{-SR} \cap m_{\mathrm{a}}\text{-SB}). \quad (8.15)$$

A mix of theory and simulation is used to determine the relevant parameters in the above equation. For the cross section and branching fraction, we use the predicted total inclusive SM Higgs production cross section of $\sigma(\mathrm{H}) = 51.04$ pb and branching fraction $\mathcal{B}(\mathrm{H} \rightarrow \gamma\gamma) = 2.27 \times 10^{-3}$ [1]. The efficiency $\varepsilon(\mathrm{H} \rightarrow \gamma\gamma,\, m_{\mathrm{a}}|m_H\text{-SR} \cap m_{\mathrm{a}}\text{-SB})$ is determined using simulated data. For simplicity, only simulated events from the gluon fusion Higgs boson production mode are used. A systematic is introduced in Sect. 8.3 to account for this approximation. The integrated luminosity $\int \mathcal{L}\,\mathrm{dt}$, corresponds to that of the yield in the denominator.

Having thus derived the shape of the $\mathrm{H} \rightarrow \gamma\gamma$ 2D-m_Γ template and its yield, expressed in terms of the fraction of the final signal region yield, we can then determine the total background shape, including both data-driven and $\mathrm{H} \rightarrow \gamma\gamma$ components, using Eq. 8.13. The total p_T re-weighted data-driven plus $\mathrm{H} \rightarrow \gamma\gamma$ background in the final signal region is then defined by taking the total background shape from the m_{a}-SR subset of Eq. 8.13 and scaling it by the estimated yield in the final signal region,

$$\begin{aligned} &\mathrm{bkg}(\mathrm{data_{rwgt}} + \mathrm{H} \rightarrow \gamma\gamma,\, m_\Gamma|m_H\text{-SR} \cap m_{\mathrm{a}}\text{-SR}) \\ &\quad \equiv N(m_\Gamma|m_H\text{-SR} \cap m_{\mathrm{a}}\text{-SR})\, p_{\mathrm{data,rwgt+H}\rightarrow\gamma\gamma}(m_\Gamma|m_H\text{-SR}), \end{aligned} \tag{8.16}$$

where, as before, $N(m_\Gamma|m_H\text{-SR} \cap m_{\mathrm{a}}\text{-SR})$ is determined using Eq. 8.7.

As a practical matter, since the simulated $\mathrm{H} \rightarrow \gamma\gamma$ samples are generated separately for each year, the $\mathrm{H} \rightarrow \gamma\gamma$ templates are derived individually for each year first, then combined into a full Run2 $\mathrm{H} \rightarrow \gamma\gamma$ template. Thereafter, they are combined combined with the data-driven component, which is always derived using the full Run2 data, in order to maximize the statistical power of the p_T re-weighting procedure.

Fit optimization The p_T re-weighting procedure, if done with limited data, can introduce additional fluctuations into the 2D-m_Γ shape. To correct for this and any other lingering mismodeling, and to provide a handle on the background model for the purposes of estimating uncertainties, we parametrize the background model obtained thus far by multiplying it with a 2-D polynomial surface,

$$\mathrm{pol}(m_{\Gamma,1}, m_{\Gamma,2}) = p_0 + p_1 m_{\Gamma,1} + p_2 m_{\Gamma,2} + \mathcal{O}(2) \tag{8.17}$$

with parameters p_i determined by a likelihood fit. Assuming as we did in Eq. 8.13 that the background shape varies proportionately over the 2D-m_Γ plane, the fit can be performed in an unbiased way by using only the m_{a}-SB region. We thus fit the parametrized background model to the observed distribution in the m_H-SR $\cap$ m_{a}-SB region, as illustrated in Fig. 8.11. We find no improvement in the goodness-of-fit beyond a polynomial of $\mathcal{O}(1)$. Additionally, checks on the impacts and extracted signal strength, similar to those performed later in Sects. 9.1.2 and 9.1.3, suggest that the analysis is insensitive to increases in the polynomial order beyond $\mathcal{O}(1)$.

The final 2D-m_Γ background model used in this analysis is that obtained after multiplying the previous background model with the above polynomial $\mathrm{pol}(m_{\Gamma,1}, m_{\Gamma,2})|_{\hat{\mathbf{p}}}$, evaluated under the under the best-fit parameters $\hat{\mathbf{p}} = \{\hat{p}_i\}$, for $i \in 0, 1, 2$, given in Table 8.1,

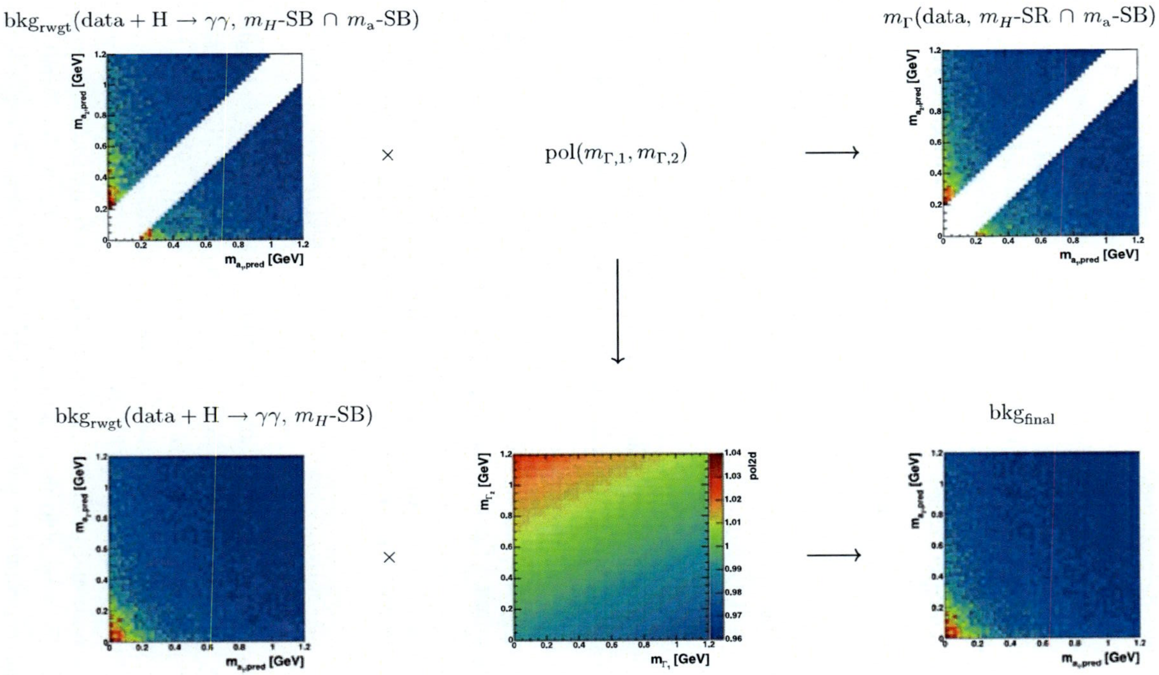

Fig. 8.11 Illustration of the 2D-m_Γ background fit optimization. Top: using the m_a off-diagonals, the combined data-driven plus $\text{H} \rightarrow \gamma\gamma$ background template, $\text{bkg}_{\text{rwgt}}(\text{data} + \text{H} \rightarrow \gamma\gamma,\ m_H\text{-SB} \cap m_a\text{-SB})$, times a 2-D polynomial $\text{pol}(m_{\Gamma,1}, m_{\Gamma,2})$ is fit against the data, $m_\Gamma(\text{data},\ m_H\text{-SR} \cap m_a\text{-SB})$, to derive the fit parameters of the polynomial surface. Bottom: the fitted $\text{pol}(m_{\Gamma,1}, m_{\Gamma,2})$ is then used to correct the full data-driven plus $\text{H} \rightarrow \gamma\gamma$ background template, $\text{bkg}_{\text{rwgt}}(\text{data} + \text{H} \rightarrow \gamma\gamma,\ m_H\text{-SB})$, to obtain the final, background model, $\text{bkg}_{\text{final}}$

Table 8.1 Best-fit parameters in pol$(m_{\Gamma,1}, m_{\Gamma,2})$ background model optimization

Fit parameter	Best-fit value	Uncertainty
p_0	9.91007×10^{-1}	6.10861×10^{-3}
p_1	-1.47380×10^{-2}	6.87198×10^{-3}
p_2	2.84057×10^{-2}	6.74935×10^{-3}
χ^2/ndf	$365/339 = 1.077$	
p-value	$0.156\ (1.0\sigma)$	

The fit parameters (left column) are shown alongside their best-fit values (middle column) and their corresponding uncertainties (right column). For reference, the reduced chi-square, χ^2/ndf, and the corresponding p-value, are also given (bottom rows)

$$\begin{aligned} \text{bkg}_{\text{final}}(m_\Gamma | m_H\text{-SR} \cap m_a\text{-SR}) \equiv \text{bkg}(\text{data}_{\text{rwgt}} + \text{H} \to \gamma\gamma,\ m_\Gamma | m_H\text{-SR} \cap m_a\text{-SR}) \\ \times \text{pol}(m_{\Gamma,1}, m_{\Gamma,2})|_{\hat{\mathbf{p}}}, \end{aligned} \tag{8.18}$$

Going forward, unless specified otherwise, references to "background model" are to be identified with the quantity $\text{bkg}_{\text{final}}$, as given above.

Summary To summarize, the full background model of the analysis is derived as follows. We assume that the 2D-m_Γ shape in the m_H-SR signal region can be expressed as a linear sum of the 2D-m_Γ shapes in the two m_H-SBs plus the simulation-derived $\text{H} \to \gamma\gamma$ 2D-m_Γ contribution in the m_H-SR. Since the p_T of the background processes affects the 2D-m_Γ and the p_T in turn varies as a function of m_H, we first re-weight the 2D-p_T in the m_H-SB sidebands to match that in the m_H-SR signal region. Having done so, we add the 2D-m_Γ shapes from the m_H-SB sidebands as described, and combine this with the simulation-derived $\text{H} \to \gamma\gamma$ 2D-m_Γ template. For the resulting combined template, we then assume that the ratio in the number of events along the 2D-m_Γ diagonal region (m_a-SR) between background and that observed in the m_H-SR is the same as that in the off-diagonal region (m_a-SB). Taking the ratio from the off-diagonal region, we then use this to normalize the background 2D-m_Γ shape in the diagonal region. This provides an unbiased estimate of the data-driven background in the final signal observation region (m_H-SR $\cap$ m_a-SR). To account for the $\text{H} \to \gamma\gamma$ contribution, we add its corresponding 2D-m_Γ shape, derived from simulation, and scale it by the fraction $f_{\text{H}\to\gamma\gamma}$, derived using theory and simulation. The data-driven background component is consequently scaled by $1 - f_{\text{H}\to\gamma\gamma}$ to preserve the overall normalization. Finally, to achieve the best possible modeling, we correct the above background model by multiplying it with the 2D polynomial obtained in the best fit of the off-diagonal (m_a-SB) region between the m_H-SB and m_H-SR.

A summary of the event yields for the different background components estimated using the above procedure is given in Table 8.2. The background components consist of the total, data-driven non-resonant background (QCD dijet, γ + jet, and prompt $\gamma\gamma$ production), and the total inclusive SM $\text{H} \to \gamma\gamma$ contribution.

Table 8.2 Estimated background yield per component, per year of data-taking

Background component	2016	2017	2018	Total RunII
Non-resonant	231,557	443,185	421,106	1,095,847
$H \rightarrow \gamma\gamma$	392	403	603	1399
Total	231,949	443,588	421,709	1,097,246

The upper row (Non-resonant) consists of the total, data-driven component (QCD dijet, γ + jet, and prompt $\gamma\gamma$ production) while the lower row ($H \rightarrow \gamma\gamma$) consists of the total inclusive SM $H \rightarrow \gamma\gamma$ component. The last column (Total RunII) lists the tally over the three years

Table 8.3 Expected signal sensitivity in the orthogonal sample space illustrating negligible signal contamination

m_a	100 MeV	400 MeV	1 GeV
N_{events}	734	698	312
$S/\sqrt{B}$	0.7	0.7	0.3
$S/(S+B)$	0.001	0.001	<0.001

The total background yield after inverting the relative track isolation is $N_{\text{bkg}} = 1{,}110{,}685$

8.2.3 *Background Validation*

To validate the background modeling procedure, we construct an orthogonal data sample that is enriched with γ + jet events, whose final signal region can be fully unblinded with negligible risk of signal contamination. This additionally allows us to validate the performance of the mass regressor in a sample that is separately photon and jet enriched. The orthogonal sample is obtained by inverting the relative track isolation requirement (nominally $(\mathcal{I}_{\text{tk}}/p_{\text{T},\Gamma}) < 0.07$) in the photon identification criteria of the subleading photon candidate Γ_2, to increase its jet content. Since this sample will also be enhanced with the p_{T}-dependent component of the QCD jet spectrum, it offers an even more stringent test of the p_{T} re-weighting procedure than is required for the nominal signal sample. We denote this orthogonal sample $\Gamma_1(\text{nom}), \Gamma_2(\text{inv})$, to distinguish it from the nominal signal sample, $\Gamma_1(\text{nom}), \Gamma_2(\text{nom})$. The signal contamination in this orthogonal sample is verified in Table 8.3.

Orthogonal sample, Γ_1(nom), Γ_2(inv). After carrying out all the background estimation steps described earlier, we obtain the expected and observed 2D-m_Γ spectra for the $\Gamma_1(\text{nom}), \Gamma_2(\text{inv})$selection in Fig. 8.12. To better compare the agreement between expected and observed distributions, we additionally plot the 2D-m_Γ distributions "unrolled": scanning along bins of increasing $m_{\Gamma,2}$ at fixed $m_{\Gamma,1}$ before incrementing in $m_{\Gamma,1}$. These are shown for the off-diagonal m_a-SB region in Fig. 8.13, and for the unblinded, diagonal m_a-SR region in Fig. 8.14.

In the upper panel of these plots, the black points (Obs) are the observed data values, with error bars corresponding to statistical uncertainties. The blue line corresponds to the background model, with the blue shaded area corresponding to its statistical uncertainties (Bkg, stat). The systematic uncertainties associated with the

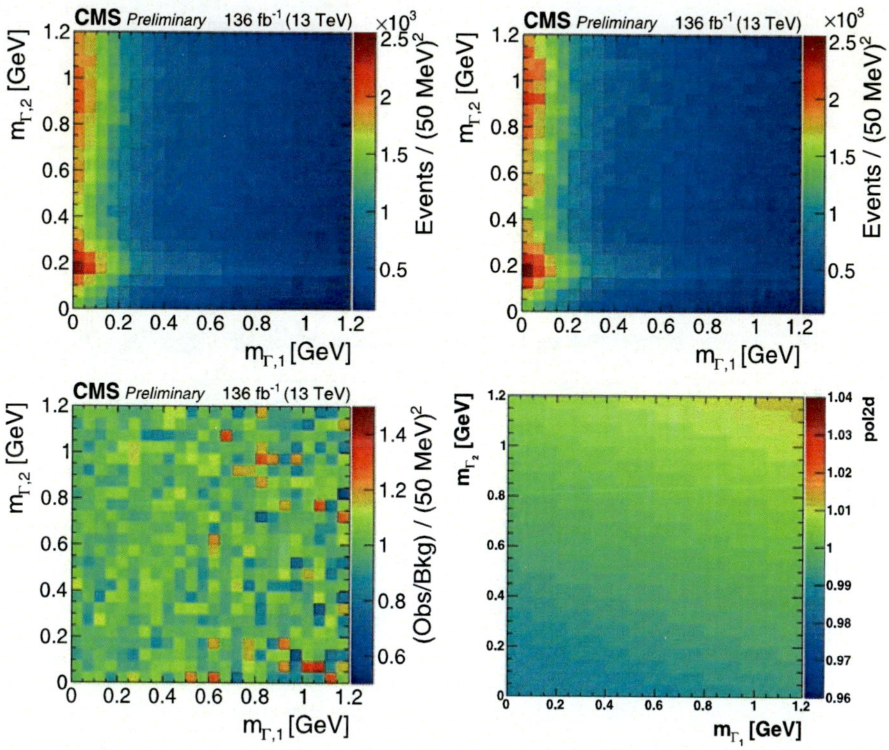

Fig. 8.12 Expected and observed 2D-m_Γ spectra for orthogonal γ + jet-enriched sample, Γ_1(nom), Γ_2(inv). Top left: observed 2D-m_Γ. Top right: expected 2D-m_Γ. Lower left: ratio of observed over expected 2D-m_Γ. Lower right: Fitted pol($m_{\Gamma,1}$, $m_{\Gamma,2}$) surface between observed and expected 2D-m_Γ.

background model, added in quadrature with the statistical uncertainties, are shown as a green band (Bkg, stat + syst). In the lower panel, the ratio of the observed over background value is shown as black points, with error bars corresponding to statistical uncertainties in the former. The ratio of the statistical plus systematic uncertainties, added in quadrature, over the background model, is shown as a green band. For maximum signal sensitivity, the distribution used as a background model in the actual maximum likelihood estimation (see Sect. 9.1) is the unrolled one.

To guide the eye with a more physically intuitive axis, comparisons for the projected 1D-m_Γ spectra of each merged photon candidate Γ are also shown, for $m_{\Gamma,1}$ (lower left plot) and $m_{\Gamma,2}$ (lower right plot). These are not used in the likelihood estimation, and are only shown for clarity. The plotting scheme is identical to that of the unrolled one, described above. Recall that the expected yields in the m_a-SB region agree with the observed ones by construction, and that this region is also used to determine the pol($m_{\Gamma,1}$, $m_{\Gamma,2}$) best-fit parameters.

We find good agreement within statistical uncertainties. Given the different views each plot has of the full 2D-m_Γ plane, by construction, we expect to see different

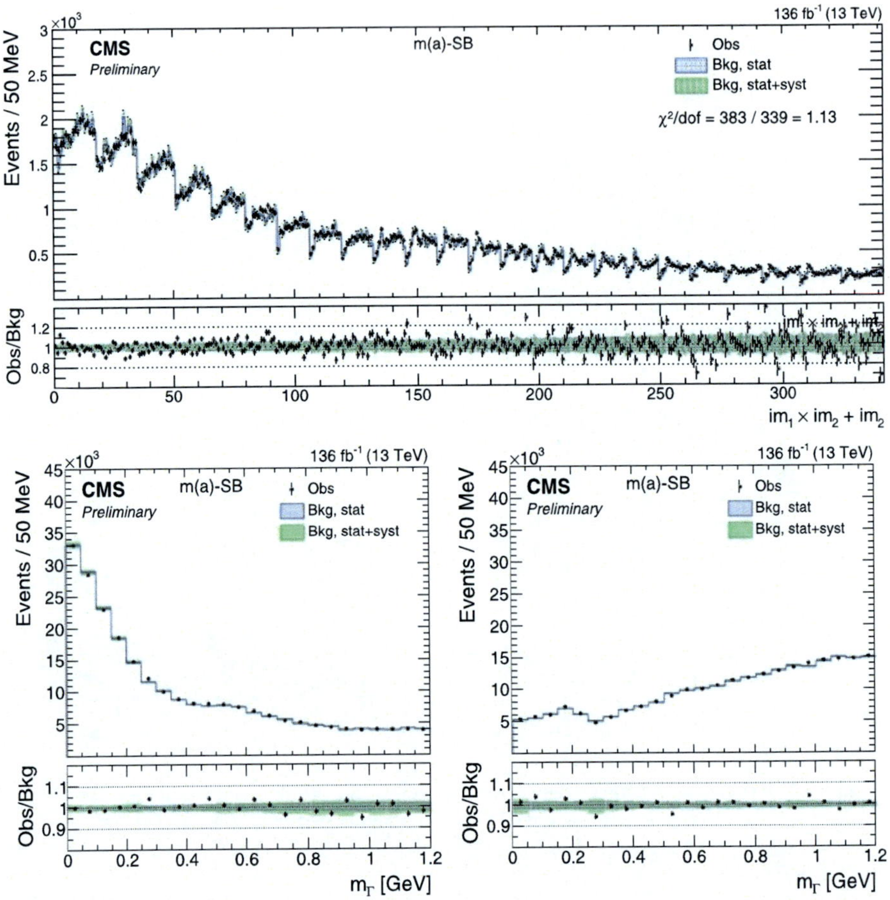

Fig. 8.13 Expected versus observed 2D-m_Γ spectra in m_a-SB for orthogonal γ + jet-enriched sample, Γ_1(nom), Γ_2(inv). Upper: unrolled 2D-m_Γ distribution. Lower: projected 1D-m_Γ distributions corresponding to the leading (left) and subleading (right) photon candidate. The event yields agree by construction. In the upper panels of each plot, the black points (Obs) are the observed data values, with error bars corresponding to statistical uncertainties. The blue line corresponds to the background model, with the blue band corresponding to its statistical uncertainties (Bkg, stat). The systematic uncertainties associated with the background model, added in quadrature with the statistical uncertainties, are shown as a green band (Bkg, stat + syst). The goodness-of-fit (χ^2/dof) between the expected and observed distributions is also displayed. In the lower panels, the ratio of the observed over background value is shown as black points, with error bars corresponding to statistical uncertainties in the former. The ratio of the statistical plus systematic uncertainties, added in quadrature, over the background model, is shown as a green band

regions of the jet spectrum in these plots. The expected enrichment in QCD jets in Γ_2 is apparent in the projected 1D-$m_{\Gamma,2}$ spectra. Specifically, in the subleading m_a-SR plot (Fig. 8.14, lower right), a π^0-like peak is visible, whereas in the m_a-SB plot (Fig. 8.13, lower right) the broad high-m_a component can be seen. Recall, however, that these features are subject to the effects of distortion from the collimation of

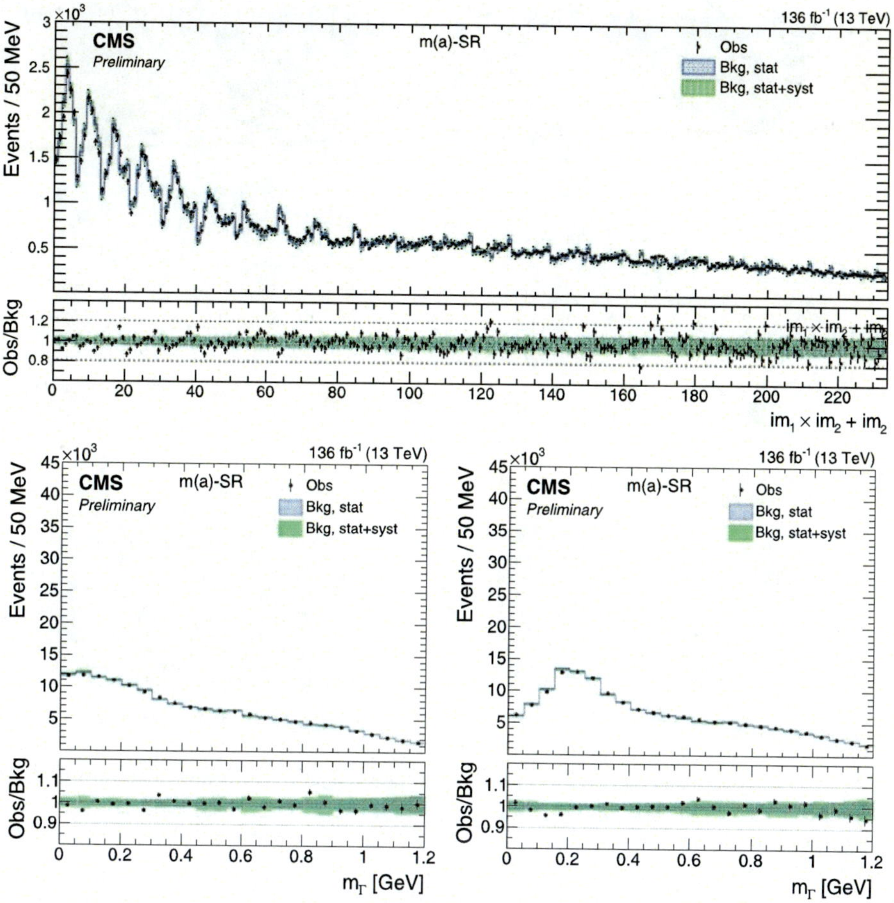

Fig. 8.14 Expected versus observed 2D-m_Γ spectra in m_a-SR for orthogonal γ + jet-enriched sample, Γ_1(nom), Γ_2(inv). Upper: unrolled 2D-m_Γ distribution. Lower: projected 1D-m_Γ distributions corresponding to the leading (left) and subleading (right) photon candidate. In the upper panels of each plot, the black points (Obs) are the observed data values, with error bars corresponding to statistical uncertainties. The blue line corresponds to the background model, with the blue band corresponding to its statistical uncertainties (Bkg, stat). The systematic uncertainties associated with the background model, added in quadrature with the statistical uncertainties, are shown as a green band (Bkg, stat + syst). In the lower panels, the ratio of the observed over background value is shown as black points, with error bars corresponding to statistical uncertainties in the former. The ratio of the statistical plus systematic uncertainties, added in quadrature, over the background model, is shown as a green band

the jet (c.f. Figs. 7.5 and 8.8). In both sets of figures, the $m_{\Gamma,1}$ spectra (lower left) appear smoothly falling, as expected for photons. These plots illustrate the reduction in background achieved by using the 2D-m_Γ spectrum as a signal discriminant: had we not directly measured m_Γ and instead measured $m_{\Gamma\Gamma}$, all the events in the m_a-SB would be directly contributing to our signal region. Lastly, the fitted pol($m_{\Gamma,1}, m_{\Gamma,2}$) surface is plotted in the lower right panel of Fig. 8.12, which shows good agreement with unity.

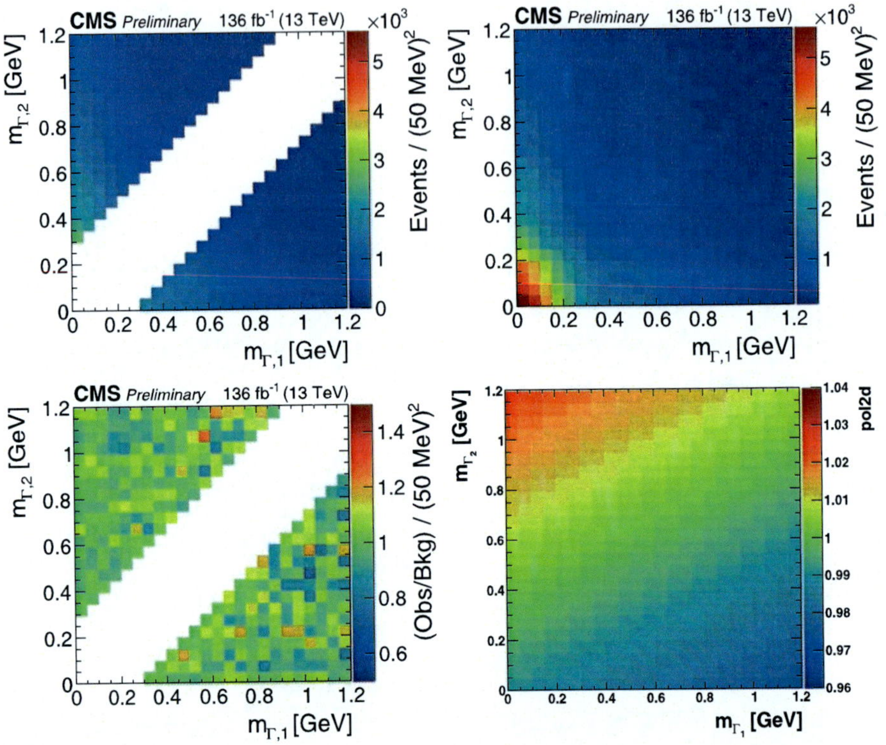

Fig. 8.15 Expected and observed 2D-m_Γ spectra for nominal signal sample, Γ_1(nom), Γ_2(nom). Top left: observed 2D-m_Γ (blinded). Top right: expected 2D-m_Γ. Lower left: ratio of observed over expected 2D-m_Γ (blinded). Lower right: Fitted pol$(m_{\Gamma,1}, m_{\Gamma,2})$ surface between observed and expected 2D-m_Γ.

Nominal signal sample, $\boldsymbol{\Gamma_1}$(nom), $\boldsymbol{\Gamma_2}$(nom). We perform a similar validation for the nominal signal sample in the m_a-SB, prior to unblinding. The expected and observed 2D-m_Γ spectra are shown in Fig. 8.15. The corresponding unrolled 2D-m_Γ, and projected 1D-m_Γ distributions in the m_a-SB and the blinded m_a-SR regions are shown in Figs. 8.16 and 8.17, respectively. The plotting scheme is as described earlier in the Γ_1(nom), Γ_2(inv)case.

In the unrolled figures, while the spectra are artificially sculpted by each region's view in the full 2D-m_Γ, the projected 1D-m_Γ spectra (lower panels) are mostly photon-like but with some contributions from isolated $\pi^0 \to \gamma\gamma$ decays (Fig. 8.17, lower right) and possibly $\eta \to \gamma\gamma$ decays (Fig. 8.16, lower left). However, for the latter, we caution that multiple merged π^0 decays contribute to a broad distribution in this vicinity as well (see discussion in Chap. 7 and Appendix A.1). That these resonant decays are not completely suppressed by the identification criteria is unavoidable as they belong to a subset of the jet phase space that bears a strong resemblance to the $a \to \gamma\gamma$ topology. Notwithstanding, their presence provides confirmation of the m_Γ regressor working as intended. The fitted pol$(m_{\Gamma,1}, m_{\Gamma,2})$ surface is shown in the

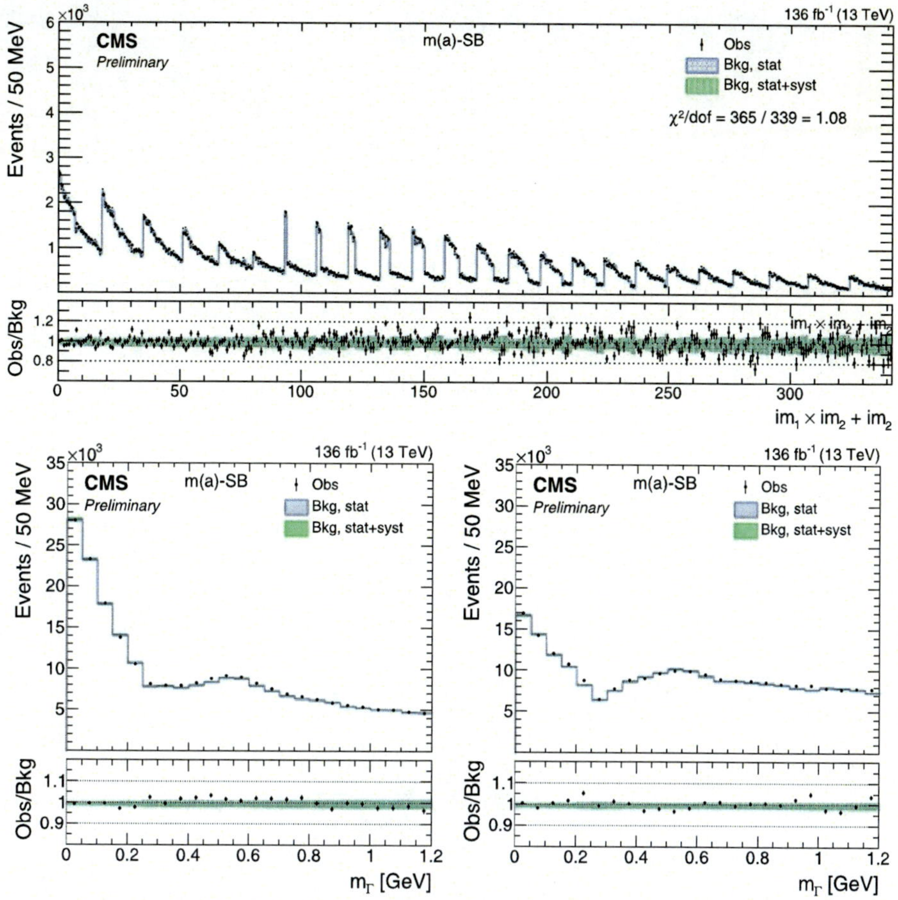

Fig. 8.16 Expected versus observed 2D-m_Γ spectra in m_a-SB for nominal signal sample, Γ_1(nom), Γ_2(nom). Upper: unrolled 2D-m_Γ distribution. Lower: projected 1D-m_Γ distributions corresponding to the leading (left) and subleading (right) photon candidate. The event yields agree by construction. In the upper panels of each plot, the black points (Obs) are the observed data values, with error bars corresponding to statistical uncertainties. The blue line corresponds to the background model, with the blue band corresponding to its statistical uncertainties (Bkg, stat). The systematic uncertainties associated with the background model, added in quadrature with the statistical uncertainties, are shown as a green band (Bkg, stat + syst). The goodness-of-fit (χ^2/dof) between the expected and observed distributions is also displayed. In the lower panels, the ratio of the observed over background value is shown as black points, with error bars corresponding to statistical uncertainties in the former. The ratio of the statistical plus systematic uncertainties, added in quadrature, over the background model, is shown as a green band

lower right panel of Fig. 8.15. We note that the agreement of the pol($m_{\Gamma,1}, m_{\Gamma,2}$) surface with unity was observed to improve with increasing background statistics. We thus attribute the mild slope to be consistent with the effects of finite statistics. These are accounted for by introducing appropriate pol($m_{\Gamma,1}, m_{\Gamma,2}$) fit uncertainties in the background model, as described in Sect. 8.3.

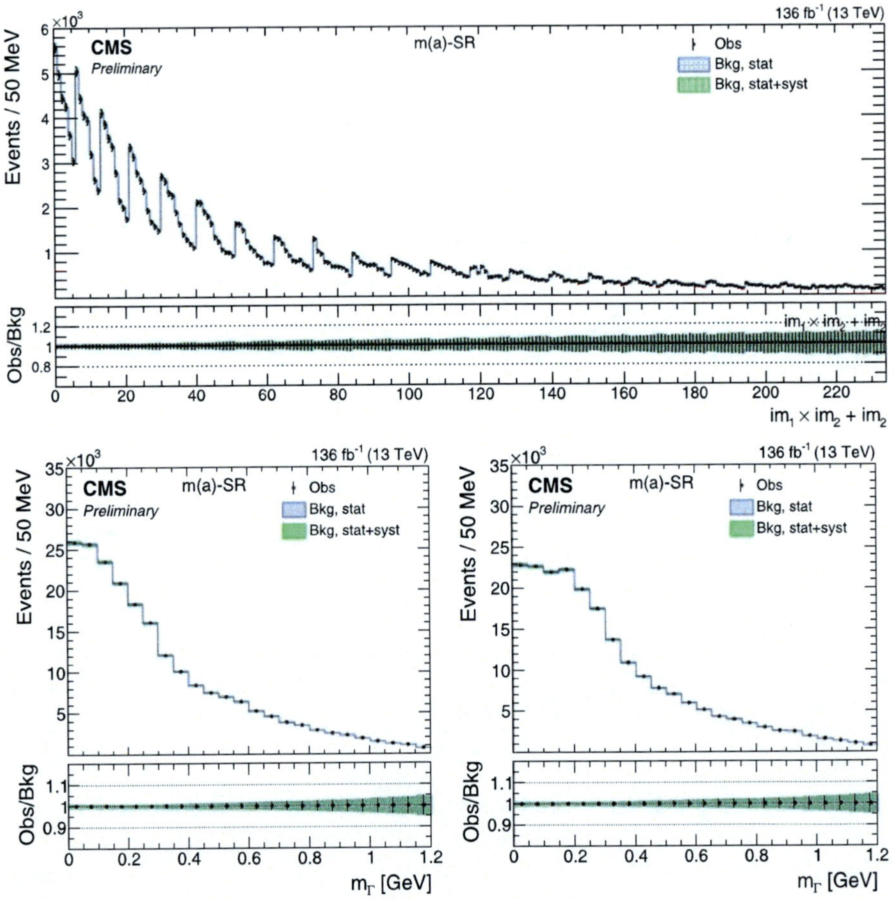

Fig. 8.17 Expected 2D-m_Γ spectra in m_a-SR for nominal signal sample, Γ_1(nom), Γ_2(nom). Upper: unrolled 2D-m_Γ distribution. Lower: projected 1D-m_Γ distributions corresponding to the leading (left) and subleading (right) photon candidate. In the upper panels of each plot, the black points (Obs) are the (blinded) observed data values, with error bars corresponding to statistical uncertainties. The blue line corresponds to the background model, with the blue band corresponding to its statistical uncertainties (Bkg, stat). The systematic uncertainties associated with the background model, added in quadrature with the statistical uncertainties, are shown as a green band (Bkg, stat + syst). In the lower panels, the ratio of the (blinded) observed over background value is shown as black points, with error bars corresponding to statistical uncertainties in the former. The ratio of the statistical plus systematic uncertainties, added in quadrature, over the background model, is shown as a green band

8.3 Systematics

The systematics associated with this analysis contribute as either shape or normalization uncertainties for either the background or signal 2D-m_Γ templates. Their various sources and effect on these templates are discussed below.

8.3.1 p_T Re-weighting

Due to the sensitivity of the m_a spectra of QCD jets to the underlying p_T, an important step in the derivation of the background model is the application of a p_T re-weighting procedure (Sect. 8.2.2) to match the 2D-p_T distribution of the source m_H-SB events from which the background is derived to that observed for the target m_H-SR events we are trying to model. Since the re-weighting procedure relies on the *total* 2D-p_T distributions rather than the QCD-only component, i.e., dijet plus γ + jet, that needs the fixing, this may potentially introduce a shape systematic on m_a if there are large discrepancies between the QCD-only and total 2D-p_T distributions. As shown in Fig. 8.18, within the limited available simulated statistics, we see no evidence of this discrepancy being realized. While our photon identification criteria already provides strong suppression of the most critical p_T-dependent jet component, it is nonetheless desirable to define a systematic that captures our confidence in the total versus QCD-only 2D-p_T agreement.

Nominally, the fraction of each m_H-SB$_{\text{low (high)}}$ in the combined 2D-p_T distribution used as input for the p_T re-weighting (see Fig. 8.9) is determined by the relative total event yields of each sideband. As calculated from Eq. 8.10 (before accounting for the H $\rightarrow \gamma\gamma$ contribution), this gives $f_{\text{SB-low,nom}} = 0.648$.

Alternatively, we choose $f_{\text{SB-low}}$ so that the relative QCD-only event yields from the m_H-SB$_{\text{low (high)}}$, as determined from simulation, are in agreement with those in the m_H-SR, modulo statistical uncertainties. If we denote the fraction of QCD events in each $m_{\Gamma\Gamma}$ data region i as $f_{\text{QCD},i}$, this amounts to the condition

$$f_{\text{QCD-low}} f_{\text{SB-low (high)}} + f_{\text{QCD-high}}(1 - f_{\text{SB-low,nom}}) = f_{\text{QCD-SR}} \pm \text{stat. uncert.}, \tag{8.19}$$

where each $f_{\text{QCD},i}$ is treated as a distribution following Poisson statistics. Solving for $f_{\text{SB-low,alt}}$ gives

$$f_{\text{SB-low (high)}} = \frac{f_{\text{QCD-SR}} - f_{\text{QCD-high}}}{f_{\text{QCD-low}} - f_{\text{QCD-high}}} \pm \text{stat. uncert.} \tag{8.20}$$

We then define the one standard deviation upper (lower) shift in the background model as the result of varying $f_{\text{SB-low}}$ by

$$f_{\text{SB-low}} = f_{\text{SB-low,nom}} \pm |f_{\text{SB-low,nom}} - f_{\text{SB-low (high)}}|. \tag{8.21}$$

To estimate the QCD event yield, because of the limited QCD dijet statistics in simulation and the relative availability of non-QCD simulated samples, we determine the QCD event yield from the difference in the total data yield minus those from non-QCD sources, as determined from simulation. The non-QCD sources include simulated prompt diphoton production and H $\rightarrow \gamma\gamma$ decays, and are normalized to the corresponding integrated luminosity of the data. A breakdown of the event yields for the major contributing processes is given in Table 8.4 for the 2017 year of data-

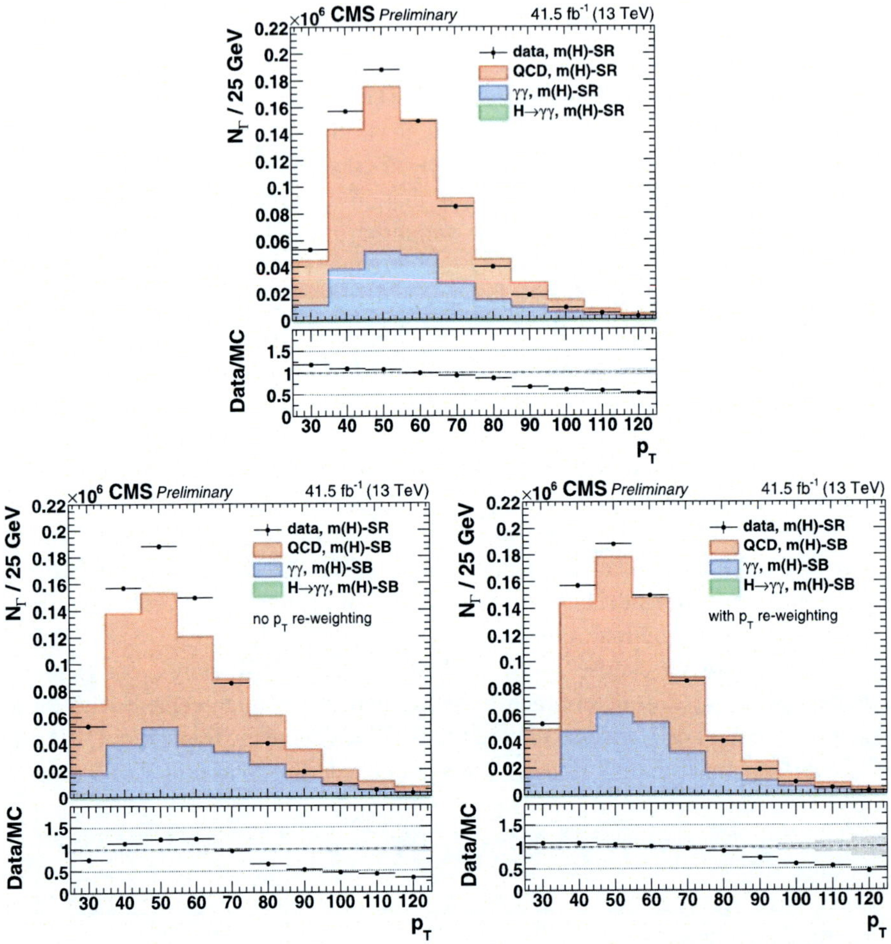

Fig. 8.18 Reconstructed p_T distributions for 2017 data versus m_H-SB-derived simulation, with and without p_T re-weighting. Top: data versus simulation, both in m_H-SR, for reference. Lower left: data in m_H-SR versus simulation in m_H-SB, without p_T re-weighting. Lower right: data in m_H-SR versus simulation in m_H-SB, with p_T re-weighting. All simulated samples normalized to integrated luminosity in data

taking. Due to missing simulated samples in the other years, only 2017 simulated samples are used to estimate this systematic for the full Run II. The lower statistics are accounted for in the statistical uncertainties, which is calculated as the sum in quadrature of those in simulation plus data.

Using this procedure, we calculate $f_{\text{SB-low}} = 0.648 \pm 0.025$, including statistical uncertainties. The size of this systematic in the background uncertainty is illustrated in Fig. 8.19 as an overlay (green band) on the background model.

Table 8.4 The data-normalized yields for simulated background events, per process, per m_H control region, for 2017 data-taking

Process	N_{events}		
	m_H-SB_{low}	m_H-SR	m_H-SB_{high}
$H \rightarrow \gamma\gamma$	6	1162	0
$\gamma\gamma$	65,192	108,746	57,852
QCD, total	191,712	245,856	85,941
γ + jet	132,445	178,341	66,320
dijet	59,267	67,516	19,622
Total	256,910	355,764	143,794

The QCD yield is further broken down into γ + jet and dijet components. *Note* Since QCD processes have the largest theoretical cross section uncertainties, only the normalization of the QCD components are adjusted to achieve overall normalization with data

8.3.2 Background Parametrization

In Sect. 8.2.2, we described how the m_H-SB-derived 2D-m_Γ background model is scaled by a polynomial pol$(m_{\Gamma,1}, m_{\Gamma,2})$ to optimize and provide a handle on the systematic uncertainties of the background modeling. To each of the best-fit polynomial parameters $\hat{p}_i, i \in [0, 2]$, is associated an uncertainty that can impact the shape of the 2D-m_Γ background template. In general, these parameter uncertainties are correlated and must be projected onto a linearly independent basis in order to obtain independent handles on the background model. In practice, this is done by calculating the covariance matrix of the fit and solving for the eigenvalues. The eigenvectors $\{\mathbf{e}_j\}$ form a linearly independent reparametrization of the fit with (uncorrelated) eigenvariations λ_j given by the square root of the eigenvalues. An uncertainty envelope in the shape of the background template with respect to each eigenvariation j can then be obtained by varying the nominal best fit vector by $\hat{\mathbf{p}} \rightarrow \hat{\mathbf{p}} \pm \lambda_j \mathbf{e_j}$ in the parametrized background model, for each eigenvariation. The resulting shift in the background model with respect to each eigenvariation is then treated as a separate shape uncertainty in the background model, with a distinct systematic introduced for each eigenvariation. The effect of the leading eigenvariation λ_0 on the shape of the 2D-m_Γ background model is shown as a green band around the nominal background shape (blue line) in Fig. 8.20.

As each eigenvariation tends to affect different bins of 2D-m_Γ to different extents, its impact on analysis sensitivity depends on the signal m_a under consideration. For instance, while λ_0 has an outsize impact on sensitivity for $m_a \approx 100$ MeV, its impact on the other m_a hypotheses is negligible.

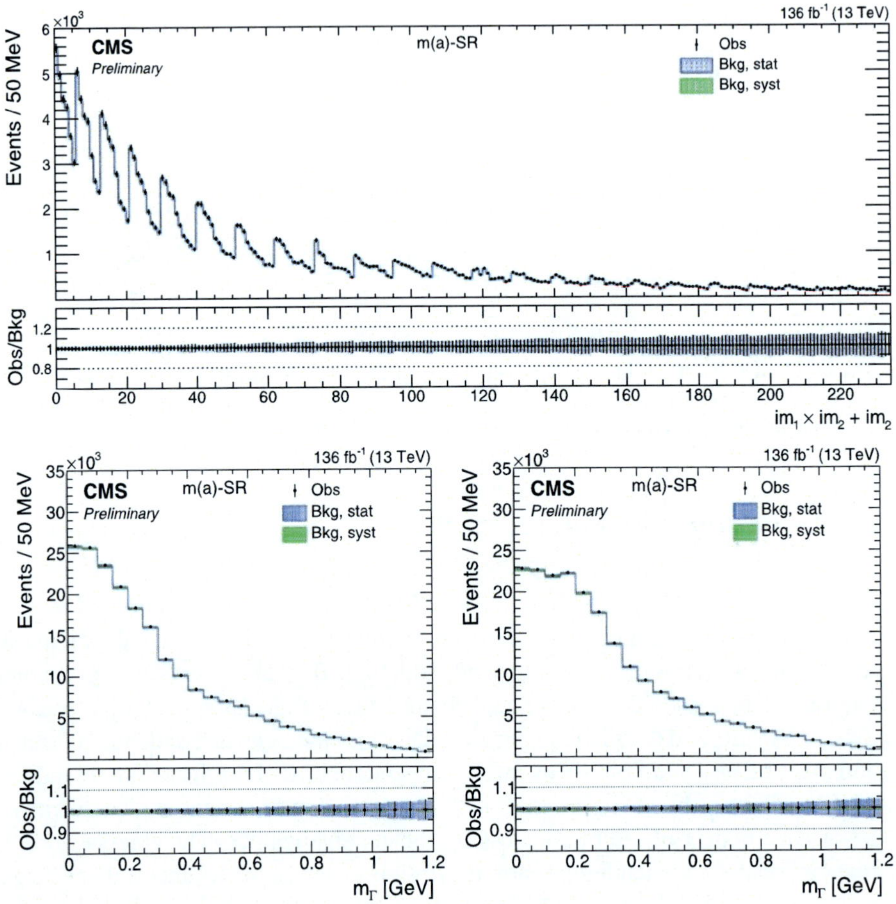

Fig. 8.19 Systematic uncertainty in the background model due to p_T re-weighting procedure in the m_a-SR. Upper: unrolled 2D-m_Γ distribution. Lower: projected 1D-m_Γ distributions corresponding to the leading (left) and subleading (right) photon candidate. In the upper panels of each plot, the black points (Obs) are the (blinded) observed data values, with error bars corresponding to statistical uncertainties. The blue line corresponds to the background model, with the blue shaded area corresponding to its statistical uncertainties (Bkg, stat). The uncertainty due to the systematic of interest is shown as a green band (Bkg, syst). In the lower panels, the ratio of the (blinded) observed over background value is shown as black points, with error bars corresponding to statistical uncertainties in the former. The ratio of the statistical plus systematic uncertainties, added in quadrature, over the background model, is shown as a green band

8.3.3 $H \rightarrow \gamma\gamma$ Template Fraction

The fraction $f_{H\rightarrow\gamma\gamma}$ of the $H \rightarrow \gamma\gamma$ 2D-m_Γ template relative to the total background template, as derived in Sect. 8.2.2, carries an uncertainty due to the parameters used to calculate it, as described in Sect. 8.2.2. While the uncertainty is associated with the yield, or normalization, of the $H \rightarrow \gamma\gamma$ 2D-m_Γ shape template, because the H

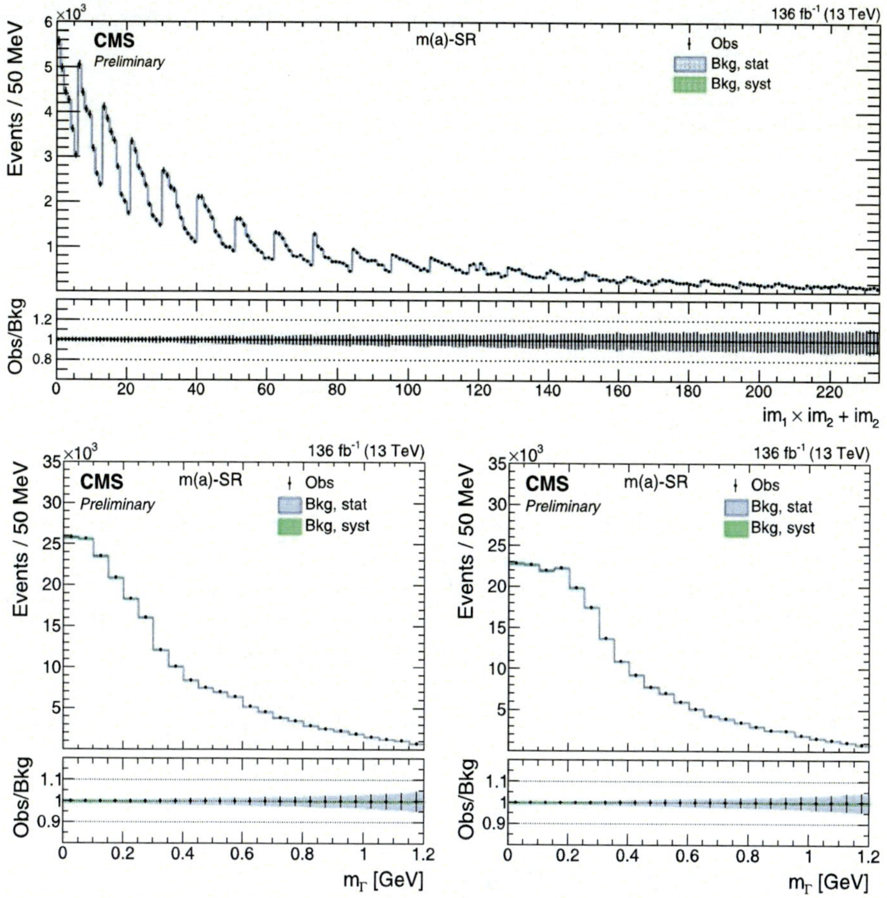

Fig. 8.20 Systematic uncertainty in the background model due to leading pol($m_{\Gamma,1}, m_{\Gamma,2}$) eigenvariation in the m_a-SR. Upper: unrolled 2D-m_Γ distribution. Lower: projected 1D-m_Γ distributions corresponding to the leading (left) and subleading (right) photon candidate. In the upper panels of each plot, the black points (Obs) are the (blinded) observed data values, with error bars corresponding to statistical uncertainties. The blue line corresponds to the background model, with the blue shaded area corresponding to its statistical uncertainties (Bkg, stat). The uncertainty due to the systematic of interest is shown as a green band (Bkg, syst). In the lower panels, the ratio of the (blinded) observed over background value is shown as black points, with error bars corresponding to statistical uncertainties in the former. The ratio of the statistical plus systematic uncertainties, added in quadrature, over the background model, is shown as a green band

$\rightarrow \gamma\gamma$ template is embedded into the total background model, in general, varying the yield of the $\mathrm{H} \rightarrow \gamma\gamma$ template introduces an uncertainty in the shape of the total background model. A shape systematic in the background model is thus required to account for this effect.

Table 8.5 Relative uncertainty in the LHC integrated luminosity affecting yield in simulated signal templates, broken down by year (columns) and type (rows).

Type	2016 (%)	2017 (%)	2018 (%)
Uncorrelated	1.0	2.0	1.5
Correlated	0.6	0.9	2.0
Correlated, 2017–2018	–	0.6	0.2

The uncertainty in the $H \to \gamma\gamma$ yield is driven by theoretical uncertainties in the total SM Higgs boson production cross section $\sigma(H) = 51.94^{+4.51}_{-4.36}$ pb [1], and the approximation used in deriving the selection efficiency $\varepsilon(H \to \gamma\gamma,\ m_a|m_H\text{-SR} \cap m_a\text{-SB})$ using simulated events from only the gluon-fusion production mode. Since the gluon fusion production mode of the Higgs boson involves no associated jets, it is likely to have the highest selection efficiency of the other production modes. A lower limit on the total $\sigma(H) \times \varepsilon(H \to \gamma\gamma,\ m_a|m_H\text{-SR} \cap m_a\text{-SB})$ thus occurs when the selection efficiencies of the other production modes are zero, or equivalently, when both cross section and efficiency are evaluated at their gluon fuson-only values. Under this simple but conservative scheme, we need only vary the value of $\sigma(H)$, upto the theoretical uncertainties on the upper end, and upto the the gluon-fusion-only value, $\sigma(H)(gg \to H) = 43.92$ pb, on the lower end. The resulting uncertainty in $f_{H\to\gamma\gamma}$ is then $(-14\%, +9\%)$, which we use to vary the background model with respect to this systematic.

8.3.4 Luminosity Uncertainty

The accuracy with which the CMS luminosity monitors can be calibrated to measure the LHC beam luminosity at the interaction point carries some uncertainty. This translates to an uncertainty in the normalization of the signal templates which are scaled to match the total integrated luminosity of the corresponding data sample, as given in Eq. 8.1. The estimated yield uncertainties in the signal template due to this effect are given in Table 8.5.

The uncertainties are broken down by run period and type, based on whether or not the underlying effects are correlated between the different periods.

8.3.5 Photon ID Scale Factor Uncertainty

As discussed in Sect. 8.1, scale factors (SFs) are introduced into the signal model to correct for the differences in selection efficiency in data versus simulation under the photon identification criteria. The tag-and-probe method used to calculate the number of passing and failing electron probes in each sample are determined using analytic fits on the $Z \to e^+e^-$ mass spectrum. To account for the uncertainty in the choice of fit functions, alternative functional forms are used. The difference in the calculated number of passing and failing electron probes using the alternate versus the nominal fit functions are then used to define a systematic uncertainty for the SFs.

The uncertainties, like the SFs, are calculated as a function of (p_T, η). In general, these uncertainties may impact the shape of the 2D-m_Γ signal template. Alternate signal models are therefore generated using SFs identically shifted up or down by one standard deviation. The size of the uncertainty in the SFs are shown as error bars on the nominal SF in the lower panel of Fig. 8.4, as a function of p_T and η.

8.3.6 HLT Trigger Scale Factor Uncertainty

Similarly, SFs are also introduced into the signal model to correct for the differences in selection efficiency in data versus simulation under the HLT diphoton trigger. This also uses the tag-and-probe method to calculate the number of passing and failing electron probes in each sample via analytic fits on the $Z \to e^+e^-$ mass spectrum. To account for the uncertainty in the choice of fit functions, alternative functional forms are also used. The difference in the calculated number of passing and failing electron probes using the alternate versus the nominal fit functions are then used to define a systematic uncertainty for the SFs.

The uncertainties, like the SFs, are calculated as a function of (p_T, η, R_9). In general, these uncertainties may impact the shape of the 2D-m_Γ signal template. Alternate signal models are therefore generated using SFs identically shifted up or down by one standard deviation. The size of the uncertainty in the SFs can be found in [6].

8.3.7 m_Γ Regressor Scale and Smearing

While the 2D-m_Γ background model is derived from data, the 2D-m_Γ signal model is derived from simulation. A shape systematic is thus needed to account for any differences in the m_Γ regressor response due to differences in how the detector response is modeled in simulation relative to data. We parametrize this difference in terms of a relative Gaussian scale and smearing shift applied to each photon candidate in each event of the 2D-m_Γ signal model, based on the photon candidate's pseudorapidity.

Using the m_Γ spectra of $Z \to e^+e^-$ electrons in data and simulation, and following the estimation procedure described in Sect. 7.5.3, we derive separate scale and smearing estimates for each year and η range as binned in: $|\eta| < 0.5$ (central), $0.5 \leq |\eta| < 1.0$ (middle), and $1.0 \leq |\eta| < 1.44$ (forward). In Table 8.6, we recount the results from that section here, for convenience.

Table 8.6 Estimated m_Γ regressor scale and smearing differences between data and simulation, derived separately for different years and η ranges

η		2016	2017	2018
$\|\eta\| < 0.5$	$\hat{s}_{\text{scale}}$	1.004	1.046	1.012
	$\hat{s}_{\text{smear}}$	6 MeV	–	2 MeV
$0.5 \leq \|\eta\| < 1.0$	$\hat{s}_{\text{scale}}$	0.978	1.032	1.018
	$\hat{s}_{\text{smear}}$	–	–	–
$1.0 \leq \|\eta\| < 1.44$	$\hat{s}_{\text{scale}}$	1.002	1.056	1.048
	$\hat{s}_{\text{smear}}$	10 MeV	–	–

Parameters are scanned in steps of $(\Delta s_{\text{scale}} = 0.002, \Delta s_{\text{smear}} = 2\,\text{MeV})$

We define a one-sided shape systematic (i.e. shift or no shift) to account for the estimated scale, and separately, smearing differences. For each systematic, the shifted distributions are constructed using the appropriate values for each year and η range and are then lumped together as a single nuisance parameter in the likelihood fit (see Sect. 9.1).

We clarify that the derived mass scale and smearings are *not* applied as corrections to the signal model and are used solely to define systematic shifts. For reference, the size of the scale-only systematic for the $m_a = 400\,\text{MeV}$ signal model is shown as an uncertainty overlay (green band) in Fig. 8.21. As noted earlier in Sect. 7.5.3, while the scale systematic uncertainties in the mass tails can be large, they are sub-dominant to the statistical uncertainties, as seen in the ratios in the lower panels of each plot in Fig. 8.21.

8.3.8 Summary

An overview of the relative size of each the above systematics in the 2D-m_Γ signal or background model is given in Table 8.7. Note that since the signal model falls off away from the mass peak, the uncertainties in these regions will tend to be erratic and eventually diverge. For clarity, only mass bins above the half-maximum of the signal mass peak are considered in the quoted signal model uncertainties.

An overview of the relative size of each the above systematics in the 2D-m_Γ signal or background model is given in Table 8.7. Code names for each systematic, as they appear in the following sections, together with a human-readable description, are also provided. Note that since the signal model falls off away from the mass peak, the uncertainties in these regions will tend to be erratic and eventually diverge. For clarity, only mass bins above the half-maximum of the signal mass peak are considered in the quoted signal model uncertainties.

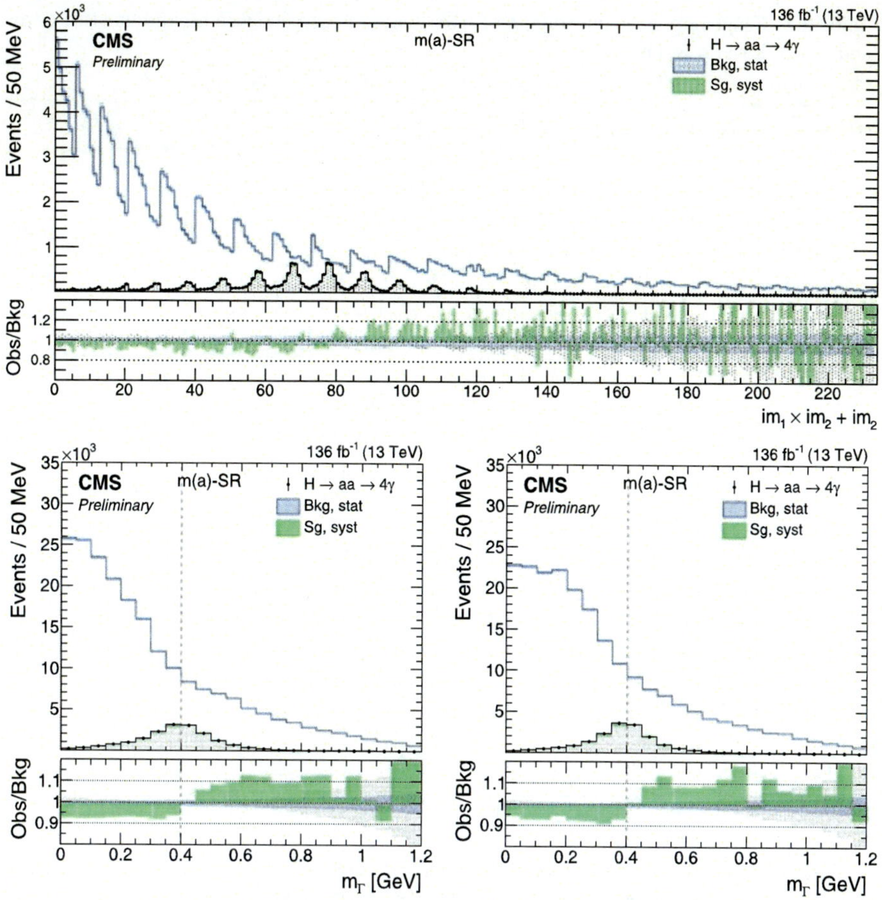

Fig. 8.21 Systematic uncertainty in the signal model due to the m_Γ regressor scale systematic in the m_a-SR. Upper: unrolled 2D-m_Γ distribution. Lower: projected 1D-m_Γ distributions corresponding to the leading (left) and subleading (right) photon candidate. In the upper panels of each plot, the black points (H $\rightarrow$ aa $\rightarrow 4\gamma$) outlining the gray region denote the signal model, with error bars corresponding to statistical uncertainties. The associated systematic uncertainties (Sg, syst) are shown as a green band. The background model is shown as a blue line, with its statistical uncertainties indicated by the blue band. In the lower panels, the ratio of the statistical (gray band) and systematic (green band) uncertainties in the signal model, over the nominal values is displayed

Table 8.7 Range of relative uncertainties for systematic uncertainties in background (upper rows) and signal (lower rows) models

Systematic	Description	Rel. uncert. (%)
prop_binh4g*	Bin-by-bin statistical uncertainties, bkg model	2–13
bgRewgtPolEigen*	pol($m_{\Gamma,1}$, $m_{\Gamma,2}$) eigenvariation λ_i for $i \in [0,2]$	<1
bgFracSBlo	Fraction of m_H-SB$_{\text{low}}$ template, $f_{\text{SB-low}}$	$\lesssim$1
bgFracHgg	Fraction of SM H $\rightarrow \gamma\gamma$ template, $f_{\text{H}\rightarrow\gamma\gamma}$	<1
prop_binh4g*	Bin-by-bin statistical uncertainties, sg model	1–4
lumi_13TeV*	LHC integrated luminosity	≈1
preselSF*	photon ID MC scale factors	≈1
hltSF*	HLT trigger MC scale factors	<1
mGamma_scale*	m_Γ regressor mass scale	$\lesssim$26
mGamma_smear*	m_Γ regressor mass smearing	<2

Left: systematic code name, as used in statistical analysis tools, relevant for the following sections. Center: human-readable description of the systematic. Right: range of relative uncertainties (Rel. Uncert.) per bin in 2D-m_Γ. For clarity in the signal systematics, only 2D-m_Γ bins with values above half-maximum are used

References

1. Heinemeyer S, Mariotti C, Passarino G, Tanaka R et al (2013). Handbook of LHC Higgs cross sections: 3. Higgs properties: report of the LHC Higgs cross section working group. CERN yellow reports: monographs, Geneva: CERN, Jul 2013. Comments: 404 pages, 139 figures, to be submitted to CERN report. Working group web page https://twiki.cern.ch/twiki/bin/view/LHCPhysics/CrossSections
2. Curtin D, Essig R, Gori S, Jaiswal P, Katz A, Liu T, Liu Z, McKeen D, Shelton J, Strassler M et al (2014) Exotic decays of the 125 GEV Higgs boson. Phys Rev D 90
3. Sirunyan AM, Tumasyan A, Adam W, Ambrogi F, Asilar E, Bergauer T, Brandstetter J, Dragicevic M, Ero J, Del Valle AE et al (2019) Combined measurements of Higgs boson couplings in proton-proton collisions at $\sqrt{s} = 13$ TeV. Eur Phys J C 79:421
4. Chatrchyan S, Khachatryan V, Sirunyan AM, Tumasyan A, Adam W, Bergauer T, Dragicevic M, Ero J, Fabjan C et al (2011) Measurement of the inclusive *w* and *z* production cross sections in pp collisions at $\sqrt{s} = 7$ TeV with the CMS experiment. J High Energy Phys 2011
5. Reconstruction of decays to merged photons using end-to-end deep learning with domain continuation in the CMS detector (2022)
6. Collaboration C (2019) CMS-AN-19-149

Chapter 9
Results

This chapter details the results of the hypothesis testing procedure comparing the compatibility of the observed data with the background-only versus signal plus background hypotheses. In Sect. 9.1, prior to unblinding the final signal region, m_H-SR $\cap$ m_{a}-SR, the analysis sensitivity is estimated in terms of the expected upper limit on excluded signal strengths, obtained by comparing the signal plus background hypothesis with the background-only one, in the final signal region m_H-SR $\cap$ m_{a}-SR. In Sect. 9.2, the unblinding is performed and the observed upper limit on excluded signal strengths is determined by comparing the signal plus background hypothesis with the actual, observed data distribution in the m_H-SR $\cap$ m_{a}-SR. The latter constitutes our final analysis result, and is expressed in terms of the branching fraction $\mathcal{B}(\mathrm{H} \to \mathrm{aa} \to 4\gamma)$ for the total signal process.

9.1 Expected Sensitivity

In this section, we estimate the sensitivity of the signal and background models, developed in the previous chapters, to the detection of a signal in data. The presence of a signal is determined using a statistical hypothesis test, where the SM background-only model (B) is taken as the null hypothesis, and the BSM signal (S) plus background model, $rS + B$, is taken as the alternative hypothesis, for some signal strength modifier or scaling parameter r. The maximum likelihood estimation (MLE) method is used to construct a test statistic for the hypothesis test by extracting the best-fit signal strength from the data in the final signal region, m_H-SR $\cap$ m_{a}-SR. In the event of an excess consistent with a signal contribution, the statistical significance of the detection is calculated. Otherwise, the $\mathrm{CL_s}$ method [1] is used to express the null result in terms of an upper limit on excluded signal strengths. To protect against potential downward fluctuations in the observed data distribution that may incorrectly fit a signal, the $\mathrm{CL_s}$ method only measures the degree of compatibility between the observed data and the $rS + B$ hypothesis, relative to the B-only hypothesis. The lower the excluded signal strength, the more stringent a constraint

M. Andrews, *Search for Exotic Higgs Boson Decays to Merged Diphotons*,
Springer Theses, https://doi.org/10.1007/978-3-031-25091-0_9

is placed on viable cross sections for the $\mathrm{H} \rightarrow \mathrm{aa} \rightarrow 4\gamma$ process. The above procedure is then repeated for each signal mass hypothesis under consideration, i.e., $m_{\mathrm{a}} = [0.1, 1]\,\mathrm{GeV}$ in 0.1 GeV steps.

Prior to unblinding the data in the final m_H-SR $\cap$ m_{a}-SR signal region, the background-only distribution in the same region is used as a proxy. In this case, the limit on the signal strength is denoted the expected upper limit, while upon using the actual observed data, it is denoted the observed upper limit. The latter constitutes the final result of this analysis. Expected and observed upper limits on signal strengths are calculated for each signal mass hypothesis (assuming no excess).

This section deals primarily with the blinded data. For simplicity, "observed data" in this section is taken to mean the background-only distribution being used as a proxy. In Sect. 9.1.1, the MLE method for constructing the hypothesis test statistic and extracting the signal strength is introduced. In Sect. 9.1.2, we asses the impact of the systematic uncertainties on the extracted signal strength. In Sect. 9.1.3, signal extraction tests are performed on various signal-depleted control regions to validate our background estimation procedure. Then, in Sect. 9.1.4, we assess the impact of the systematic uncertainties on the final analysis sensitivity. Finally, the expected analysis is summarized in Sect. 9.1.5. The results for the actual unblinded data are presented in the following this chapter.

9.1.1 Maximum Likelihood Estimation (MLE)

The primary framework used in this analysis for the detection of a signal is that of the statistical hypothesis test [2]. The SM background-only (B) model is interpreted as the null hypothesis while the BSM signal-plus-background ($rS + B$) model is interpreted as the alternative hypothesis, for some unknown signal strength scaling parameter or modifier r, representing the production cross section of the signal process. A prescription for testing whether the observed data favors one model or the other is provided by the MLE method [1]. In the MLE approach, the test statistic used to discriminate between competing hypotheses is based on the ratio of the likelihoods of the given observation under the competing hypotheses, which, per the Neyman-Pearson lemma, is the most powerful discriminant that can be used. The test statistic is then maximized with respect to the signal strength modifier to determine the value most compatible with data.

The first step in the MLE process is the construction of the likelihood function. Since this analysis is a counting experiment over several bins i in 2D-m_Γ, a so-called "shape" analysis, the likelihood is modelled as the product of Poisson probabilities over the bins i, $\mathcal{P}(n_i|rS_i(\boldsymbol{\theta} + B_i(\boldsymbol{\theta}))$, for observing n_i events in bin i, given the $\mathrm{H} \rightarrow \mathrm{aa} \rightarrow 4\gamma$ predicted yields rS_i plus the expected SM-only yields B_i. To account for the systematic uncertainties that affect our knowledge of these yields, for each systematic j, a constraint is introduced into the likelihood that allows the signal and background yields to deviate from their nominal values, subject to some cost on the likelihood. Specifically, a penalty term in the form of a normal or log-normal distribution $p(\tilde{\theta}_j|\theta_j, \sigma_j)$ is introduced for each systematic. A Bayesian interpretation for these penalty terms is followed, where the systematic strength modifier θ_j regulating the cost is associated with the distribution's mean, evaluated at $\tilde{\theta}_j = 0$, with standard deviation $\sigma_j = 1$. The addition of systematic uncertainties thus introduces auxiliary parameters into the likelihood, known as nuisance parameters (NPs), over which the likelihood will also need to be maximized in order to determine the best-fit signal strength. The likelihood function modified in this way is known as the profile likelihood and takes the form

$$\mathcal{L}(\mathbf{n}|r, \boldsymbol{\theta}) = \prod_{i \in \text{bins}} \mathcal{P}(n_i|rS_i(\boldsymbol{\theta}) + B_i(\boldsymbol{\theta})) \times \prod_{j \in \text{systs}} p(0|\theta_j, 1) \quad (9.1)$$

We note that bin-by-bin statistical uncertainties are treated as nuisance parameters in the same way that systematic uncertainties are. They differ only in the set of bins which they affect (individual versus all bins, respectively). For this reason, in what follows, "systematic uncertainties" are to be understood as including both statistical and systematic uncertainties, unless they are otherwise distinguished.

Having defined the likelihood function, the next step in the MLE process is to construct the hypothesis test statistic. Following the Neyman-Pearson lemma, the likelihood ratio-based test statistic $\tilde{q}_r$, evaluated for a given r, is given by

$$\tilde{q}_r = -2 \ln \left[\frac{\mathcal{L}(\mathbf{n}|r, \hat{\boldsymbol{\theta}}_r)}{\mathcal{L}(\mathbf{n}|\hat{r}, \hat{\boldsymbol{\theta}})} \right] \quad 0 \leq \hat{r} \leq r \quad (9.2)$$

where the numerator $\mathcal{L}(\mathbf{n}|r, \hat{\boldsymbol{\theta}}_r)$ is the likelihood in Eq. 9.1 conditioned on the given r and evaluated at its maximum (i.e. at the best-fit $\hat{\theta}_r$), and the denominator $\mathcal{L}(\mathbf{n}|\hat{r}, \hat{\boldsymbol{\theta}})$ is the likelihood of Eq. 9.1 evaluated at its unconditioned maximum (i.e. at the best-fit $(\hat{r}, \hat{\boldsymbol{\theta}})$). To perform the hypothesis test, one needs to know the probability distribution of possible test statistic outcomes for a given r, $f(\tilde{q}_r|r)$. Determining this distribution, however, requires running multiple pseudo-experiments on toy models that sample the $\{\theta_j\}$ parameter space of the profile likelihood in Eq. 9.1, and repeating this for every r to be studies. Fortunately, in the limit of large bin counts[1], as is the case in this analysis, $f(\tilde{q}_r|r)$ approaches a χ^2 distribution, as given by Wilk's theorem. Asymptotic formulas thus exist to quickly and analytically determine $f(\tilde{q}_r|r)$, as implemented in the Higgs combine software package [1].

In the event of a non-zero best-fit signal strength, to quantify the significance of the signal, the value of the resulting test statistic, $\tilde{q}_{\hat{r}}$, is compared with the probability distribution of the test statistic under the background-only hypothesis, $f(\tilde{q}_0|0)$. In particular, the right-side tail probability $\text{P}(\tilde{q}_0 > \tilde{q}_{\hat{r}}|0)$ is calculated, otherwise known as the p-value for the observation given the background-only hypothesis, p-val(obs$|B$). The p-value is thus to be understood as the probability of obtaining a

[1] Specifically, in the limit the Poisson distribution modeling the bin counts can be approximated as a Gaussian distribution, which underlies the bin counts of the χ^2 distribution.

test static value more deviant (greater) than that observed in the data, assuming the B-only hypothesis. As a matter of convention, this p-val is often recast in terms of the tail probabilities of an equivalent standard normal distribution, giving the famed "$N\sigma$" criteria of detection significance.

For any best-fit signal strengths with a detection significance less than 3σ, the result is interpreted as a statistical fluctuation consistent with the SM B-only hypothesis. In this scenario, the non-detection is interpreted as an upper limit on signal strengths ruled out by the analysis, following the $\mathrm{CL_s}$ prescription [1]. In addition to p-val(obs$|B$), the $\mathrm{CL_s}$ method requires scanning over different signal strengths hypotheses and determining the corresponding p-values, p-val(obs$|rS + B$). The $\mathrm{CL_s}$ metric is then given by

$$\mathrm{CL_s} = \frac{\mathrm{p} - \mathrm{val}(\mathrm{obs}|rS + B)}{1 - \mathrm{p} - \mathrm{val}(\mathrm{obs}|B)} \tag{9.3}$$

While strictly not a p-value itself, the $\mathrm{CL_s}$ metric is nonetheless used in a similar way to define confidence levels of excluded signal strengths. From the scan in r, signal strengths $r > r_\alpha$ such that $\mathrm{CL_s} < \alpha$, are said to be excluded by the data upto a confidence level of $1 - \alpha$. That is, r_α is the upper limit on signal strengths excluded at the $1 - \alpha$ confidence level. In particular, the excluded signal strength at the 95% confidence level is denoted $r_{0.05}$.

Reporting exclusion limits, however, as done in later sections, requires taking the above procedure a step further. In order to report the *median* expected upper limit and its corresponding $\pm 1, 2\sigma$ confidence intervals—the so-called Brazilian plots—the probability distribution *over* $r_{0.05}$ needs to be determined. This requires running pseudo-experiments on toy data sampled from the background-only hypothesis, and calculating $\mathrm{CL_s}$ and $r_{0.05}$ again for each. Once this probability distribution over $r_{0.05}$ is obtained, the median expected upper limit is defined as the $r_{0.05}$ with 50% cumulative probability, and the 1 and 2σ confidence intervals, those with 68% and 95% cumulative probability, respectively. These confidence intervals are not to be confused with the confidence levels used as $\mathrm{CL_s}$ thresholds in the individual pseudo-experiments. For simplicity, since only the median expected upper limit is ever reported, going forward, "expected upper limit" or "$r_{0.05}$" is to always be understood as the median expected upper limit on excluded signal strengths at the 95% confidence level.

As opposed to traditional p-value formulations, the $\mathrm{CL_s}$ approach has two main advantages that let it err on the side of caution. As noted earlier in the section, it protects the analysis from a potential downward fluctuations in the observed data distribution that may incorrectly fit a signal, and make a false discovery. By the same vein, it also protects the analysis from excluding signal strengths for which it has no sensitivity to distinguish from B-only.

Pior to unblinding, the MLE method is primarily used to define a set of metrics and tests for evaluating the expected analysis sensitivity, given the signal and background models and their uncertainties. In the following subsections, we study the impact of the model uncertainties on the extracted signal strengths, and later, the median expected upper limit on excluded signal strengths. The final expected exclusion limits are given at the end of this section.

9.1.2 Impacts

We now calculate a few instructive metrics to evaluate the effect each systematic uncertainty has on the best-fit signal strength $\hat{r}$ from the likelihood fit. These are calculated by assuming a signal strength modifier of $r = 1$, to reveal any biases the systematics might introduce into the likelihood fit—and thus the extracted best-fit $\hat{r}$—in the presence of a signal. We then run pseudo-experiments: using MC methods, we generate observed distributions corresponding to the background model but with input systematic strength modifiers $\{\theta_j\}$ randomly sampled from their respective distributions $p(0|\theta_j, 1)$[2], or so-called Asimov data sets. For each pseudo-experiment, we re-fit the likelihood for the best-fit signal strength $\hat{r}$ and best-fit systematic strengths $\hat{\theta}_j$, and use these to populate impact and pull distributions, respectively, corresponding to the differences between the best-fit value and the input value.

The pull and impact distributions are shown below for the leading thirty systematics by impact, for the $m_a = 100$ MeV (Fig. 9.1), 400 MeV (Fig. 9.2), and $m_a = 1$ GeV (Fig. 9.3) signal hypotheses. As a matter of convention, in all impact plots, the signal models at each mass hypothesis have been normalized to the median expected upper limit at the 95% confidence level, $r_{0.05}$, for that mass hypothesis. A dictionary for the naming convention used in these plots is provided in Table 8.7.

At the $m_a \approx 100$ MeV mass hypothesis (Fig. 9.1), the largest impact to the best-fit signal strength (right panel) comes from the leading pol$(m_{\Gamma,1}, m_{\Gamma,2})$ eigenvariation λ_0 (top row), roughly corresponding to the uncertainty in the overall normalization of the background model. Since this low-m_a region of the 2D-m_Γ plane has the highest background contamination, even a small uncertainty in the background normalization here can have an outsize effect on the measured signal strength. Indeed, from Table 8.7, the relative uncertainty in any one individual 2D-m_Γ bin due to variations in any of the $pol(m_{\Gamma,1}, m_{\Gamma,2})$ parameters is $< 1\%$. Compounding this effect, however, is the fact the $m_a \approx 100$ MeV mass resonance is the most difficult to reconstruct. As seen in Figs. 8.1–8.3, the mass peak at this regime is at its broadest and, for the same signal cross section, is much smaller than the background yield in its vicinity. Since the $pol(m_{\Gamma,1}, m_{\Gamma,2})$ uncertainties act as "catch-alls" for any lingering, residual background mismodeling, their magnitude and ranking in the impacts (first and third) are consistent with what we expect for a data-driven background model amid a challenging signal. The second-ranking impact, the fraction $f_{\text{SB-low}}$ associated with the p_T re-weighting procedure, is also data-driven. This suggests that there are no particularly dominant systematics affecting the background model, other than the statistical fluctuations in the data-driven workflows of the background estimation. The remaining rankings in the impact at $m_a = 100$ MeV are the independent, bin-by-bin statistical uncertainties.

In terms of the pulls (left panel), we observe no significant shifts in the pull distributions that would suggest a systematic is biasing the likelihood fit. We do note that the statistical uncertainties are all uniformly slightly under-constrained. We attribute this to the correlation with the p_T re-weighting systematic $f_{\text{SB-low}}$. Since the

[2] Computationally, this is equivalent to just sampling from $p(\theta_j|0, 1)$.

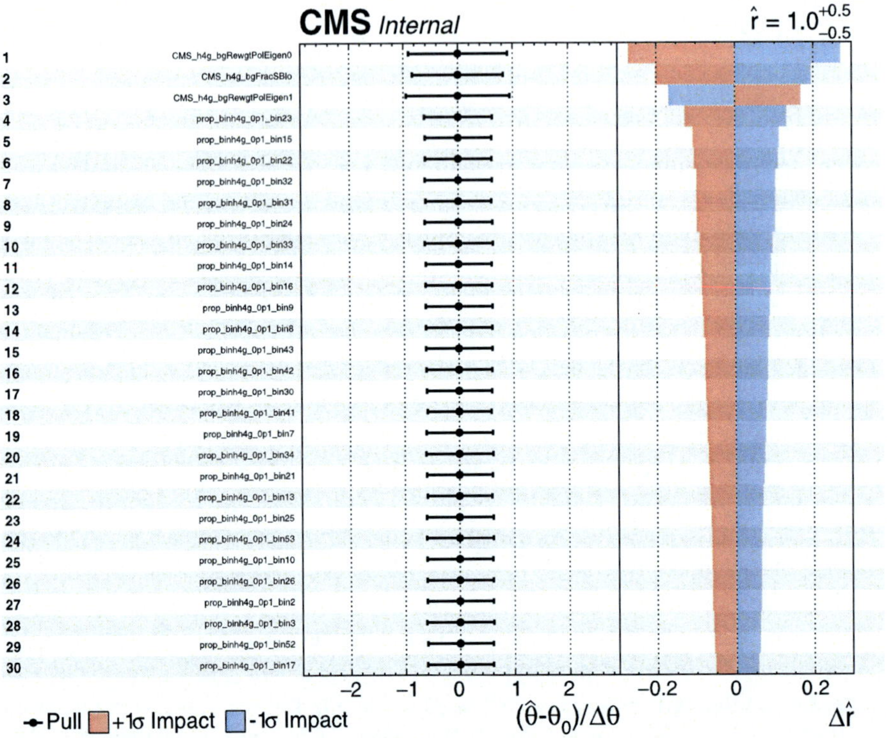

Fig. 9.1 Impact of the top thirty systematics on the best-fit signal strength (right panel) for likelihood fits with the $m_a = 100\,\text{MeV}$ signal model. The corresponding pull distributions are shown in the left panel. The signal model is normalized to the median expected upper limit at 95% confidence, $r_{0.05}$, for this mass hypothesis. The naming convention follows the one given in Table 8.7, but with `CMS_h4g_` prepended

choice of f_{SB-low} affects the weights with which events in the m_H-SBs are used to fill the background model, we expect the resulting bin-by-bin statistical uncertainties in the background model to be correlated with variations in $f_{\text{SB-low}}$. While not visible in the top 30 impacts, as noted in Sect. 8.3.7, the m_Γ regressor scale and smearing uncertainties are one-sided, by construction, and thus their impacts are one-sided as well.

The impact plots for the remaining mass hypotheses (Figs. 9.2 and 9.3) are dominated by the individual bin-by-bin statistical uncertainties. We attribute the less dramatic impact of the $pol(m_{\Gamma,1}, m_{\Gamma,2})$-induced uncertainties to the relatively stronger significance and yield of the signal relative to background at these mass regimes. The broad conclusion, however, is the same, namely, that the uncertainty in the best-fit signal strength is dominated by statistical uncertainties, either indirectly through the data-driven workflows of the background estimation, or directly from the bin-wise statistical uncertainties. We do not observe any shifts in the pull distributions at these m_a hypotheses either.

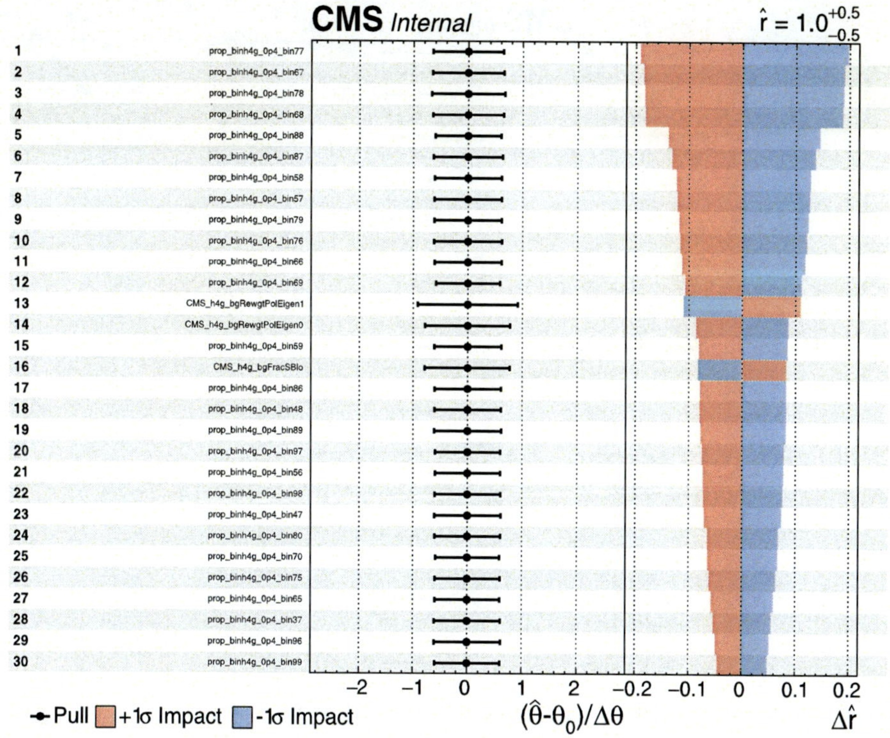

Fig. 9.2 Impact of the top thirty systematics on the best-fit signal strength (right panel) for likelihood fits with the $m_a = 400$ MeV signal model. The corresponding pull distributions are shown in the left panel. The signal model is normalized to the median expected upper limit at 95% confidence, $r_{0.05}$, for this mass hypothesis. The naming convention follows the one given in Table 8.7, but with `CMS_h4g_` prepended

9.1.3 Control Region Signal Extraction Tests

To validate the entire analysis chain, we perform a full signal extraction test on various unblinded control regions of the data. Since we expect these regions to be signal-depleted, the measurement of a statistically significant signal at any mass hypothesis in these regions would be an indication of a potential bias in one or more of the underlying components of the analysis. In particular, such a check can test the quality of the background modeling: if a correlated and sustained significance above 3σ is observed between neighboring mass hypotheses, this may suggest a potential weakness in the background model. It is worth noting, however, that, due purely to statistical fluctuations, crossings above 3σ are to be expected, even with a well-modeled background, a phenomenon known as the look elsewhere effect [3].

The results of the extracted signal significance are shown in Fig. 9.4. Three validation regions are chosen, utilizing the available unblinded regions in the m_a-SB of the nominal selection (Γ_1(nom), Γ_2(nom), see Sect. 8.2.3) and the m_a-SB or m_a-SR

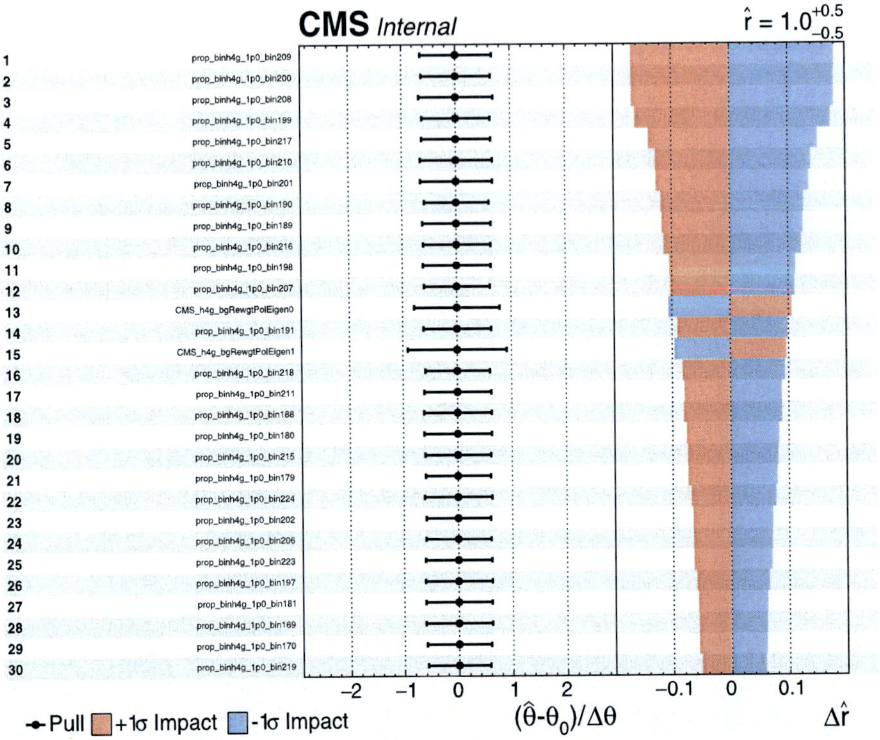

Fig. 9.3 Impact of the top thirty systematics on the best-fit signal strength (right panel) for likelihood fits with the $m_a = 1$ GeV signal model. The corresponding pull distributions are shown in the left panel. The signal model is normalized to the median expected upper limit at 95% confidence, $r_{0.05}$, for this mass hypothesis. The naming convention follows the one given in Table 8.7, but with `CMS_h4g_` prepended

of the inverted selection (Γ_1(nom), Γ_2(inv), see Sect. 8.2.3), using the unrolled 2D-m_Γ. In all three cases, the signal model from the final signal region, the m_a-SR with the nominal selection, is used, otherwise no signal peak would be present. The top plot shows the extracted signal significance for the observed distribution with the nominal selection in the m_a-SB versus the background model of the same plus signal model of the nominal selection in the m_a-SR. The middle plot corresponds to the observed distribution with the inverted selection in the m_a-SB versus the background model of the same plus signal model of the nominal selection in the m_a-SR. In both cases, we see no statistically significant excess in the observed distribution over the background. One caveat in these two plots, however, is that the unrolled 2D-m_Γ of the signal model, which comes from the m_a-SR rather than the m_a-SB that the observed and background distributions come from, is directly and naively overlaid onto the latter distributions. As such the signal bins do not correspond to the same physical bins in the original 2D-m_Γ plane, and the bin alignment between the distributions is entirely artificial. Of course, only so much can be done without unblinding the final signal region. Alternatively, we consider the observed versus background distributions in the m_a-SR of the inverted selection, even though these correspond to an orthogonal selection, with, again, the signal model from the nominal selection in the m_a-SR. In this case, the bin alignment between the distributions is physically valid, even though the underlying event selections differ. These are shown in the bottom plot, in which we see no statistically significant signal contribution as well.

9.1.4 N-1 Expected Upper Limits

To complement the study of the impact of each systematic on the best-fit signal strength $\hat{r}$, we perform a study of the impact that each systematic ultimately propagates to the expected upper limit on the signal strength, the primary metric of analysis sensitivity. For each m_a hypothesis, we alternately remove one systematic at a time, and for each, re-calculate the expected upper limit on the signal strength $r_{0.05}$. The fractional change in the expected upper limit for each $N-1$ systematic scenario, relative to the nominal N-systematics case ($r_{0.05}$), is then calculated. The results are shown in Table 9.1. While the impact of the statistical category appears to dominate the N-1 expected upper limits across the m_a hypotheses, to be precise, each bin in 2D-m_Γ must be treated as an independent systematic and thus the category represents the effect of several bin-by-bin uncertainties, each of which is in fact sub-dominant to any one of the other systematics, as suggested by the previous impact plots in Figs. 9.1, 9.2 and 9.3. The results are strongly consistent with our findings from the impact studies. For all the mass hypotheses, the expected upper limits are most sen-

Table 9.1 The relative change in the median expected upper limit on the signal strength ($\Delta\hat{r}/\hat{r}_0$), from the removal of a single systematic uncertainty (N-1 systematic, listed on each row) in the MLE, for various m_a hypotheses. The top rows list systematics affecting the background model, while the lowers rows, those affecting the signal model

N-1 systematic	m_a		
	100 MeV	400 MeV	1 GeV
Bin-by-bin statistical uncertainties.	0.24	0.37	0.37
$pol(m_{\Gamma,1}, m_{\Gamma,2})$ eixgenvariation λ_0	0.15	0.01	0.02
$pol(m_{\Gamma,1}, m_{\Gamma,2})$ eigenvariation λ_1	0.05	0.03	0.02
$pol(m_{\Gamma,1}, m_{\Gamma,2})$ eigenvariation λ_2	–	–	–
Fraction of $m_H - SB_{low}$ template, $f_{\text{SB-low}}$	0.07	0.01	–
Fraction of SM H $\rightarrow \gamma\gamma$ template, $f_{\text{H}\rightarrow\gamma\gamma}$	–	–	–
LHC integrated luminosity	–	–	–
photon ID MC scale factors	–	–	–
HLT trigger MC scale factors	–	–	–
m_Γ regressor mass scale	–	–	–
m_Γ regressor mass smearing	–	–	–

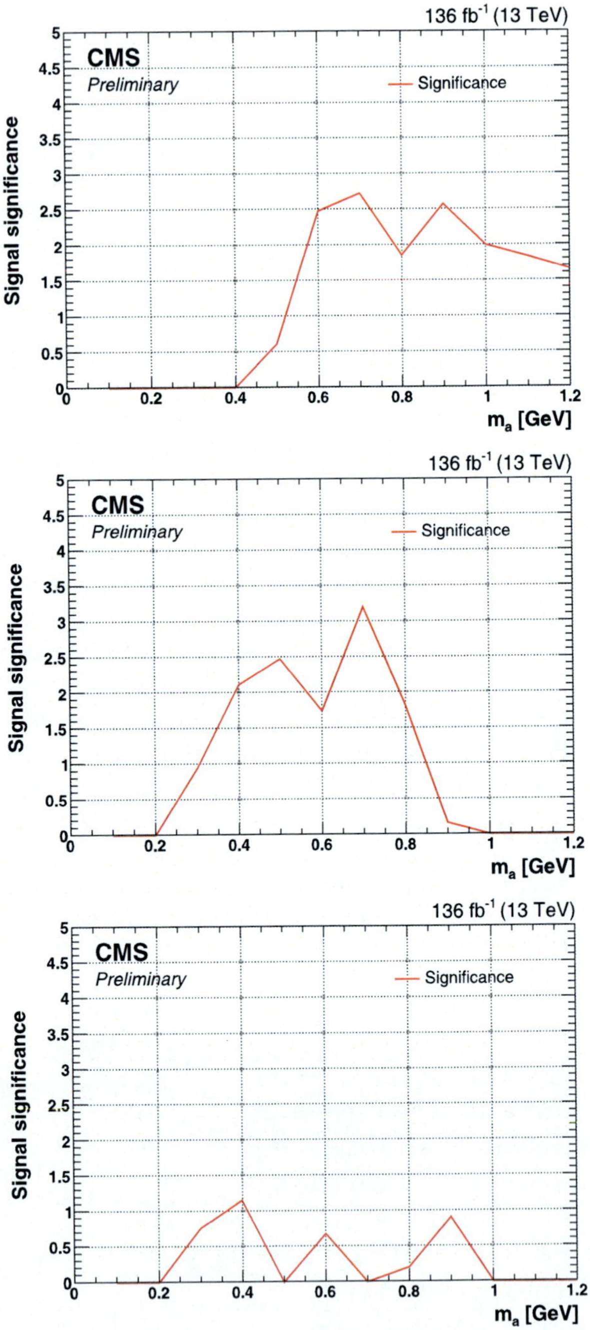

Fig. 9.4 Results of the signal significance extraction in the validation regions as a function of m_a hypothesis. Top: observed distribution with the nominal selection (Γ_1(nom), Γ_2(nom)) in the m_a-SB versus the background model of the same plus signal model of the nominal selection in the m_a-SR. Middle: observed distribution with the inverted selection (Γ_1(nom), Γ_2(inv)) in the m_a-SB versus the background model of the same plus signal model of the nominal selection in the m_a-SR. Bottom: observed distribution with the inverted selection (Γ_1(nom), Γ_2(inv)) in the m_a-SR versus the background model of the same plus signal model of the nominal selection in the m_a-SR

sitive to statistical uncertainties, followed by systematics uncertainties representing data-driven workflows. As was the case in the impact studies, the magnitude of the decrease in sensitivity varies with the relative significance and yield of the signal versus the background at a particular mass regime. Notably, despite the m_Γ regressor scale and smearing systematics showing large bin-by-bin uncertainties in Table 8.7, their effect on the expected upper limit is ultimately negligible. We thus conclude that the analysis sensitivity is primarily constrained by the effects of finite data, either directly in the form of bin-by-bin statistical uncertainties, or indirectly via uncertainties in the data-driven background estimation workflows. Improving the mass resolution of the m_Γ regressor, especially at low-m_a, may offer some opportunity to minimize these effects.

9.1.5 Expected Upper Limits

We summarize the results of the previous study of the effect of the systematics on the median expected upper limit at the 95% confidence level by presenting these over the full considered mass range of $m_a = [0.1, 1.2]$ GeV, in 0.1 GeV steps. These are shown in Fig. 9.5, with limits expressed in terms of the total $\text{H} \to \text{aa} \to 4\gamma$ signal branching fraction, $\mathcal{B}(\text{H} \to \text{aa} \to 4\gamma)$. In addition to the median expected value (blue dashed curve), the $\pm 1, 2\sigma$ confidence intervals (green and yellow bands, respectively) are also displayed, calculated using the same CL_s method. For comparison, the 95%CLcorresponding to the CMS SM $\text{H} \to \gamma\gamma$ branching fraction measurement [4], $\mathcal{B}(\text{H} \to \gamma\gamma) = (2.27 \times 10^3) \times 1.20^{+0.18}_{-0.14}$, is also plotted (red band). It is relevant for values of $m_a \approx 0.1$ GeVwhere the $\text{a} \to \gamma\gamma$ decay resembles a single photon. At larger m_a where the $\text{a} \to \gamma\gamma$ decay appears more distinct from a single photon, the SM $\text{H} \to \gamma\gamma$ event selection [5] is significantly less efficient at accepting $\text{a} \to \gamma\gamma$ decays. Based on the differences in event selection efficiency, the corresponding $\text{H} \to \gamma\gamma$ upper limit at $m_a \approx 1$ GeV is estimated to be about twice that at $m_a \approx 0.1$ GeV.

The analysis is expected to be most sensitive at around $m_a \approx 400$ MeV, gradually weakening as one approaches either side of the $m_a \in (0.1, 1.2)$ GeV range. This trend is the result of an interplay between competing effects influencing the signal significance as a function of m_a. The first is the size of the background contribution, primarily from photons, which, as shown in Fig. 8.17, is largest at $m_\Gamma \approx 0$, and falls off by nearly an order of magnitude toward higher m_Γ. The second is the event selection efficiency. The criteria in the diphoton trigger (see Chap. 6), originally optimized for photon selection, give the highest selection efficiency for highly-merged low-m_a decays, but considerably poorer efficiency for high-m_a decays. This is exacerbated by the underestimation of the p_T and the $m_{\Gamma\Gamma}$, causing further losses to accumulate under the event selection criteria. Finally, there is the m_Γ resolution reconstructed by the m_Γ regressor. At low-m_a, where the $\text{a} \to \gamma\gamma$ is most boosted and most difficult to reconstruct, the relative mass resolution is at its poorest (see Sect. 7.3). The relative resolution then improves with increasing m_a as the $\text{a} \to \gamma\gamma$ opening angle

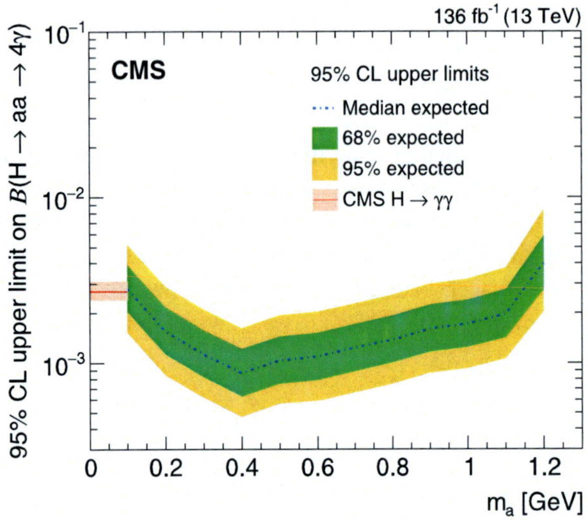

Fig. 9.5 Median expected (blue dashed curve) upper limits on excluded $\mathrm{H} \to \mathrm{aa} \to 4\gamma$ signal strengths, at the 95% confidence level (CL). The limits are expressed in terms of the total $\mathrm{H} \to \mathrm{aa} \to 4\gamma$ branching fraction $\mathcal{B}(\mathrm{H} \to \mathrm{aa} \to 4\gamma)$ and evaluated for the signal mass hypotheses $m_{\mathrm{a}} = [0.1, 1.2]\,\mathrm{GeV}$, in 0.1 GeV steps. The 68% (±1 std. deviation) and 95% (± 2 std. deviation) confidence intervals around the median expected upper limits are shown as green and yellow bands, respectively. The 95% CLupper limit corresponding to the CMS SM $\mathcal{B}(\mathrm{H} \to \gamma\gamma)$ measurement [4] is also shown for comparison (red band, where the width represents the uncertainty in the measurement)

increases, and begins to plateau around $m_{\mathrm{a}} \gtrsim 400\,\mathrm{MeV}$. The signal significance at $m_{\mathrm{a}} \approx 400\,\mathrm{MeV}$ thus occupies an optimal point where these competing effect conspire to give the best signal significance. From there, the signal significance weakens to either side as one effect or another begins to dominate over the others.

The median values of the expected upper limits on excluded signal strengths (dashed line) correspond to branching fractions in the range of $\mathcal{B}(\mathrm{H} \to \mathrm{aa} \to 4\gamma) \approx (0.8,\, 4) \times 10^{-3}$ over the m_{a} hypotheses, comparable to, or smaller than, the observed SM $\mathrm{H} \to \gamma\gamma$ branching fraction of $\mathcal{B}(\mathrm{H} \to \gamma\gamma) = 2.27 \times 10^{-3}$. As noted in Sect. 8.1, the $\mathrm{H} \to \gamma\gamma$ coupling measurement [4], insofar as its event selection criteria overlap with ours, sets an effective upper bound on allowed $\mathrm{H} \to \mathrm{aa} \to 4\gamma$ contributions at low-m_{a}, where the $\mathrm{H} \to \mathrm{aa} \to 4\gamma$ resonance would be largely buried with the SM $\mathrm{H} \to \gamma\gamma$ resonance. At $m_{\mathrm{a}} \gtrsim 400\,\mathrm{MeV}$, where the $\mathrm{a} \to \gamma\gamma$ decay is more likely to have been excluded by the standard $\mathrm{H} \to \gamma\gamma$ photon identification criteria, less measurements exist to constrain a $\mathrm{H} \to \mathrm{aa} \to 4\gamma$ contribution. Our sensitivity, therefore, is competitive with realistic values likely still allowed by existing constraints. We are also the only analysis able to set constraints for $\mathrm{H} \to \mathrm{aa}$ production below the dimuon production threshold ($m_{\mathrm{a}} \lesssim 250\,\mathrm{MeV}$). Assuming $\mathcal{B}(\mathrm{a} \to \gamma\gamma) \sim 1$, our limits are, over all considered m_{a}, an order-of-magnitude more sensitive than conservative theoretical estimates for the $\mathrm{H} \to \mathrm{aa}$ production cross section [6].

9.2 Observed Results

In this section, the data in the final signal region is unblinded and we present our final results and offer our interpretation. In Sect. 9.2.1, we present the unblinded counterpart of the studies assessing the impact of the systematics on the extracted best-fit signal strength. We then perform goodness-of-fit tests between the unblinded data and the best-fit $\hat{r}S + B$ model in Sect. 9.2.2. In Sect. 9.2.3, the unblinded data distribution in the m_H-SR $\cap$ m_{a}-SR is shown against the background distribution with the best-fit parameters from the likelihood fit. The results of the search are then summarized in terms of upper limits on excluded signal strengths in Sect. 9.2.4. A discussion of how robust these results are in light of assumptions made in the search is then given in Sect. 9.2.5.

9.2.1 *Impacts*

The pull and impact distributions in the final signal region are shown for the leading thirty systematics by impact, for the $m_{\mathrm{a}} = 100\,\mathrm{MeV}$ (Fig. 9.6), 400 MeV (Fig. 9.7), and $m_{\mathrm{a}} = 1\,\mathrm{GeV}$ (Fig. 9.8) signal hypotheses. As a matter of convention, in all impact plots, the signal models at each mass hypothesis have been normalized to the median expected upper limit at the 95% confidence level, $r_{0.05}$, for that mass hypothesis. The extracted signal strength is given in the top right. A dictionary for the naming convention used in these plots is provided in Table 8.7.

The impact rankings are consistent with those expected from the Asimov data set in Sect. 9.1.2. We see no evidence of statistically significant pulls (left panel) for any of the systematics. For all but the $m_{\mathrm{a}} = 400\,\mathrm{MeV}$ signal hypothesis, we find the best-fit signal strength to be consistent with no signal and, indeed, a downward statistical fluctuation in the data relative to the background. For $m_{\mathrm{a}} = 400\,\mathrm{MeV}$ the best-fit signal strength gives $\hat{r} = 1$, or about 2σ from the background-only hypothesis, consistent with an upward fluctuation in the data relative to the background. Since the signal models have been normalized to the expected upper limit, we expect the above deviations in the best-fit signal strength to manifest as equally large deviations in the observed upper limit relative to the expected upper limit, as we later show in Sect. 9.2.4.

Since we assumed in our background model that the contribution from the predicted SM $\mathcal{B}(\mathrm{H} \rightarrow \gamma\gamma)$ consists entirely of $\mathrm{H} \rightarrow \gamma\gamma$, if a fraction of the measured $\mathcal{B}(\mathrm{H} \rightarrow \gamma\gamma)$ were to, in fact, contain $\mathrm{H} \rightarrow \mathrm{aa} \rightarrow 4\gamma$, we would observe a downward pull in the fraction of the $\mathrm{H} \rightarrow \gamma\gamma$ component, $f_{\mathrm{H}\rightarrow\gamma\gamma}$, to accommodate the non-zero $\mathcal{B}(\mathrm{H} \rightarrow \mathrm{aa} \rightarrow 4\gamma)$. In Figs. 9.9, 9.10 and 9.11, we thus check that no such significant pull is observed for the $f_{\mathrm{H}\rightarrow\gamma\gamma}$ systematic in the unblinded impacts. Indeed, we find the pulls to be consistent with zero for this systematic.

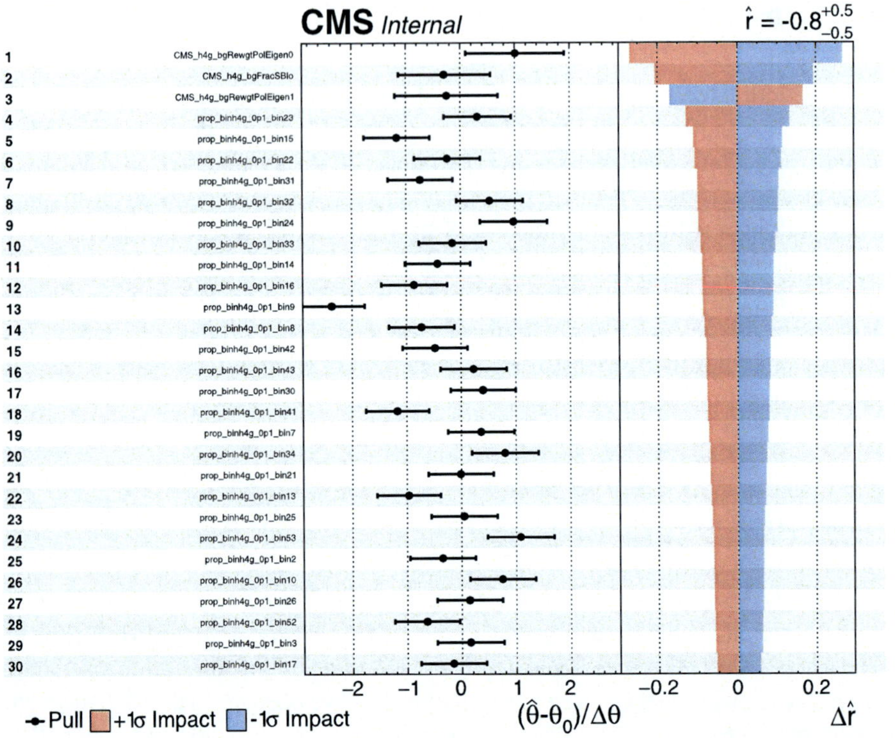

Fig. 9.6 Unblinded impact of the top thirty systematics on the best-fit signal strength (right panel) for likelihood fits with the $m_a = 100\,\text{MeV}$ signal model. The corresponding pull distributions are shown in the left panel. The signal model is normalized to the median expected upper limit at 95% confidence, $r_{0.05}$, for this mass hypothesis. The naming convention follows the one given in Table 8.7, but with `CMS_h4g_` prepended

9.2.2 *Goodness-of-fit*

Before looking at the unblinded 2D-m_Γ distributions in the final signal region, we first check the goodness-of-fit (GoF) between some observed GoF statistic for the unblinded data under the best-fit $\hat{r}S + B$ hypothesis, and a distribution of toy GoF statistics under the B-only hypothesis. The GoF statistic is given by the saturated algorithm, which is a generalization of the χ^2 test to distributions whose bin counts are non-Gaussian (i.e., for low bin counts). Pseudo-experiments over 1k toys are used to populate the test statistic distribution over the B-only hypothesis. The test reports the p-value for the observation under the background-only hypothesis which can be expressed in terms of the tail probabilities of a comparable standard normal distribution to determine the significance of the observation. The results of these comparison are shown in Fig. 9.12. The p-values are between 0.10–0.14 (modulo statistical uncertainties) corresponding to a compatibility significance of around 1.1–1.2σ. These are within acceptable limits.

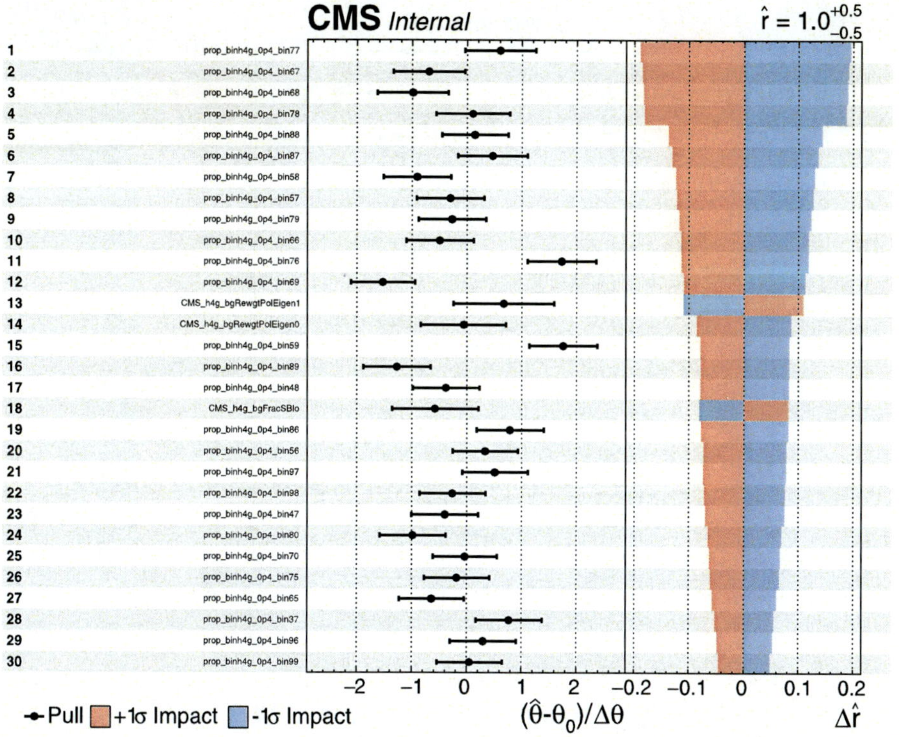

Fig. 9.7 Unblinded impact of the top thirty systematics on the best-fit signal strength (right panel) for likelihood fits with the $m_a = 400$ MeV signal model. The corresponding pull distributions are shown in the left panel. The signal model is normalized to the median expected upper limit at 95% confidence, $r_{0.05}$, for this mass hypothesis. The naming convention follows the one given in Table 8.7, but with CMS_h4g_ prepended

9.2.3 *Observed 2D-m_Γ Distribution*

The unblinded 2D-m_Γ distribution in the final signal region is shown in Fig. 9.13 for the unrolled 2D-m_Γ distribution (upper) and for the 1D-m_Γ projections corresponding to the leading (lower left) and subleading (lower right) photon candidate. The black points (Obs) are the observed data values, with error bars corresponding to statistical uncertainties. The expected background (blue line) uses the optimized values for the systematics strength modifiers determined from the likelihood fit. The total statistical plus systematic uncertainties, also determined from the likelihood fit, are shown as a green band around the background model (Bkg, stat+syst). In the lower panel, the ratio of the observed over background value is shown as black points, with error bars corresponding to statistical uncertainties in the former. The ratio of the total uncertainty over the background model is shown as a green band.

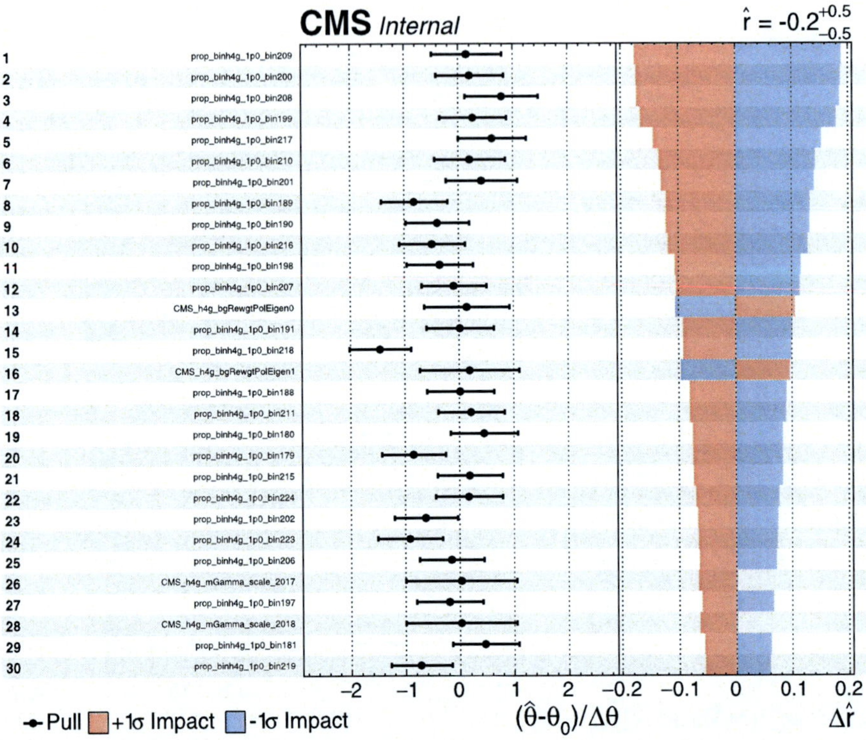

Fig. 9.8 Unblinded impact of the top thirty systematics on the best-fit signal strength (right panel) for likelihood fits with the $m_{\mathrm{a}} = 1\,\mathrm{GeV}$ signal model. The corresponding pull distributions are shown in the left panel. The signal model is normalized to the median expected upper limit at 95% confidence, $r_{0.05}$, for this mass hypothesis. The naming convention follows the one given in Table 8.7, but with `CMS_h4g_` prepended

Looking at the ratio panel in the projection of the subleading photon candidate (lower right) one can indeed make out the downward ($m_{\mathrm{a}} \approx 0.1,\ 1\,\mathrm{GeV}$) and upward ($m_{\mathrm{a}} \approx 0.4\,\mathrm{GeV}$) fluctuations in the data relative to the background, consistent with our findings in the previous Sect. 9.2.1.

9.2.4 Observed Upper Limits

The observed upper limits on the excluded signal strengths at the 95% confidence level are displayed as a solid black curve with points in Fig. 9.14. The values of the limits are expressed in terms of the total $\mathrm{H} \to \mathrm{aa} \to 4\gamma$ branching fraction $\mathcal{B}(\mathrm{H} \to \mathrm{aa} \to 4\gamma)$. The corresponding median expected upper limit (blue dashed curve), along with its $\pm 1\sigma$ (green band) and $\pm 2\sigma$ (yellow band) confidence intervals, are also shown. For comparison, the 95% CLcorresponding to the CMS SM $\mathrm{H} \to \gamma\gamma$ branching fraction measurement [4], $\mathcal{B}(\mathrm{H} \to \gamma\gamma) = (2.27 \times 10^{3}) \times 1.20^{+0.18}_{-0.14}$, is

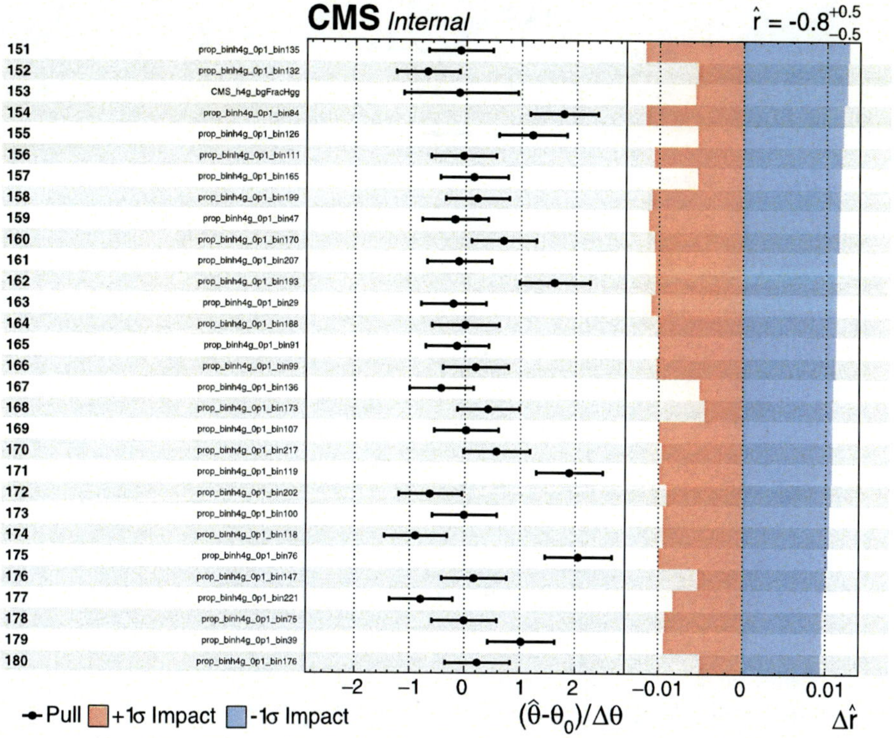

Fig. 9.9 Unblinded impact of the $f_{H\to\gamma\gamma}$ systematic (rank 153) on the best-fit signal strength (right panel) for likelihood fits with the $m_a = 100$ MeV signal model. The corresponding pull distributions are shown in the left panel. The signal model is normalized to the median expected upper limit at 95% confidence, $r_{0.05}$, for this mass hypothesis. The naming convention follows the one given in Table 8.7, but with `CMS_h4g_` prepended

also plotted (red band). As noted at the start of Sect. 9.1.5, this upper limit is relevant for values of $m_a \approx 0.1$ GeVwhere the $a \to \gamma\gamma$ decay resembles a single photon. At larger m_a where the $a \to \gamma\gamma$ decay appears more distinct from a single photon, the SM $H \to \gamma\gamma$ event selection [5] is significantly less efficient at accepting $a \to \gamma\gamma$ decays. Based on the differences in event selection efficiency, the corresponding $H \to \gamma\gamma$ upper limit at $m_a \approx 1$ GeV is estimated to be about twice that at $m_a \approx 0.1$ GeV.

We find no statistically significant $H \to aa \to 4\gamma$ signal contributions in the CMS Run II data set. We exclude models of $H \to aa \to 4\gamma$ production predicting branching fractions between $\mathcal{B} \gtrsim (0.9, 3.3) \times 10^{-3}$, depending on $m_a \in [0.1, 1.2]$ GeV, at the 95% confidence level.

The deviations in the observed versus expected upper limits are consistent with those given in the best-fit signal strengths of Figs. 9.6, 9.7 and 9.8. These are below the conventional 3σ significance threshold for evidence of new physics and are consistent with statistical fluctuations in the SM-only hypothesis. A complementary way of interpreting these deviations is to express the best-fit signal strengths at each

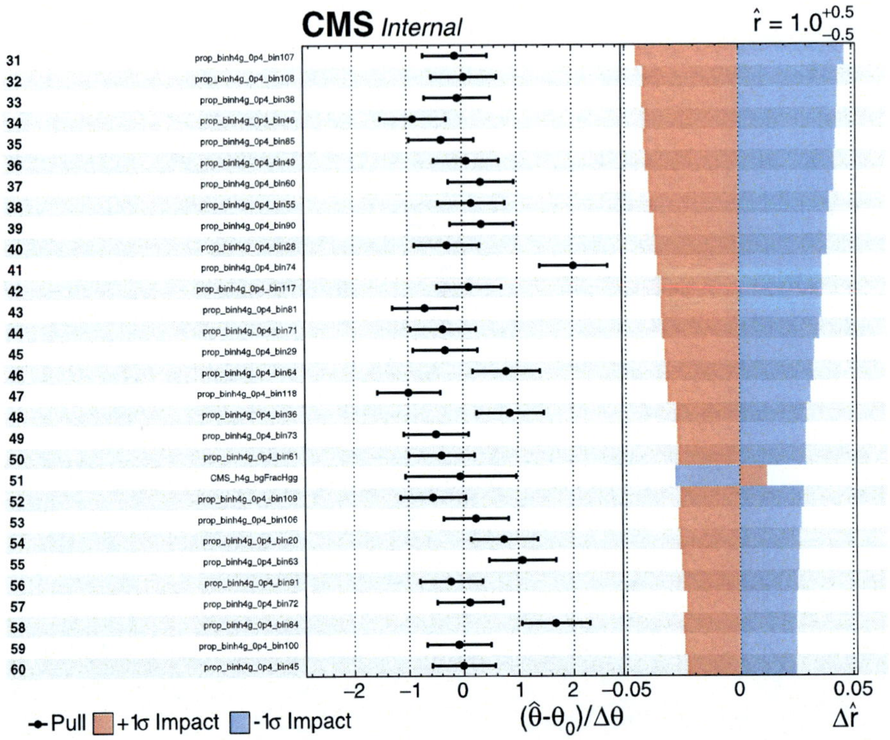

Fig. 9.10 Unblinded impact of the $f_{H\rightarrow\gamma\gamma}$ systematic (rank 51) on the best-fit signal strength (right panel) for likelihood fits with the $m_a = 400$ MeV signal model. The corresponding pull distributions are shown in the left panel. The signal model is normalized to the median expected upper limit at 95% confidence, $r_{0.05}$, for this mass hypothesis. The naming convention follows the one given in Table 8.7, but with `CMS_h4g_` prepended

mass hypothesis in terms of their detection significance. As shown in Fig. 9.15, no statistically significant excess is found for any of the m_a hypotheses. The largest excess is found at $m_a = 0.4$ GeV, at a level of 2σ.

Our observed limits are comparable to, or better than, the measured branching fraction for $H \rightarrow \gamma\gamma$ [4]. Since the $H \rightarrow aa \rightarrow 4\gamma$ signal, if realized in nature, would have a $m_{\Gamma\Gamma}$ resonance approximately degenerate with that of the SM $H \rightarrow \gamma\gamma$, the $\mathcal{B}(H \rightarrow \gamma\gamma)$ measurement constitutes an effective upper bound on a low mass $H \rightarrow aa \rightarrow 4\gamma$ contribution, insofar as our event selection agrees with theirs. In practice, our photon identification criteria are slightly looser. Since the higher $m_a \gtrsim 400$ MeV signal hypotheses have $a \rightarrow \gamma\gamma$ topologies which are more likely to fail the standard SM $H \rightarrow \gamma\gamma$ selection, constraints on these masses from the existing $\mathcal{B}(H \rightarrow \gamma\gamma)$ measurement are less conclusive. In any case, our results represent an improvement in the $\mathcal{B}(H \rightarrow aa \rightarrow 4\gamma)$ limits over existing indirect CMS constraints, for most of the mass range $m_a \in (0.1, 1.2)$ GeV. Among other $H \rightarrow aa$ decay modes, our results constitute the only dedicated constraints for masses below the dimuon threshold, $m_a \lesssim 210$ MeV.

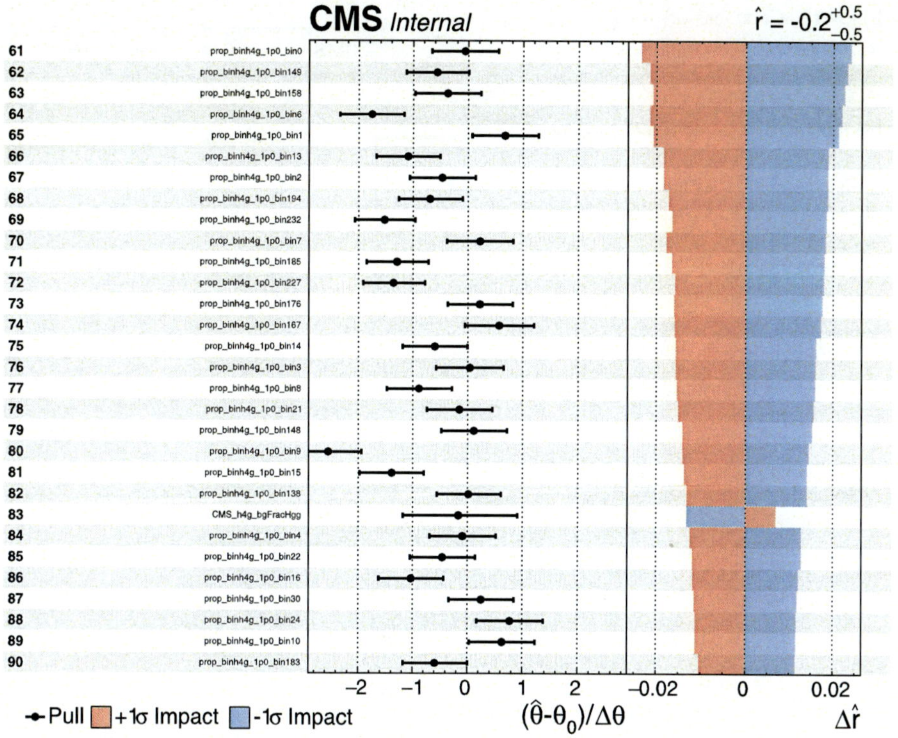

Fig. 9.11 Unblinded impact of the $f_{\mathrm{H}\to\gamma\gamma}$ systematic (rank 83) on the best-fit signal strength (right panel) for likelihood fits with the $m_{\mathrm{a}} = 1$ GeV signal model. The corresponding pull distributions are shown in the left panel. The signal model is normalized to the median expected upper limit at 95% confidence, $r_{0.05}$, for this mass hypothesis. The naming convention follows the one given in Table 8.7, but with `CMS_h4g_` prepended

The only other direct constraint on $\mathrm{H} \to \mathrm{aa} \to 4\gamma$ production at our mass regime comes from a Run I result from ATLAS [9]. While the ATLAS result was performed on a much smaller data set, despite a being an indirect search in $m_{\Gamma\Gamma}$, their results are only 30–90% worse. This primarily has to do with the finer segmentation of the ATLAS ECAL, allowing for much stronger suppression of background processes than can be achieved with the CMS ECAL, at least with current photon ID. Nonetheless, our results represent the most stringent limits on low-mass $\mathrm{H} \to \mathrm{aa} \to 4\gamma$ production thus far at the LHC.

The closest $\mathrm{a} \to \gamma\gamma$ constraints from non-$\mathrm{H} \to \mathrm{aa}$ production come from the GlueX experiment [10]. While their limits cannot be directly compared with ours, we note that, because neutral meson decays are well-isolated in their energy regime, their sensitivity is dominated by π^0 and η backgrounds, within the range of their respective masses. This is thus a distinct advantage of our analysis and mass reconstruction strategy, and the higher energy regime of the LHC, where the neutral meson decays are collimated and are easier to suppress over the isolated $\mathrm{a} \to \gamma\gamma$ decay.

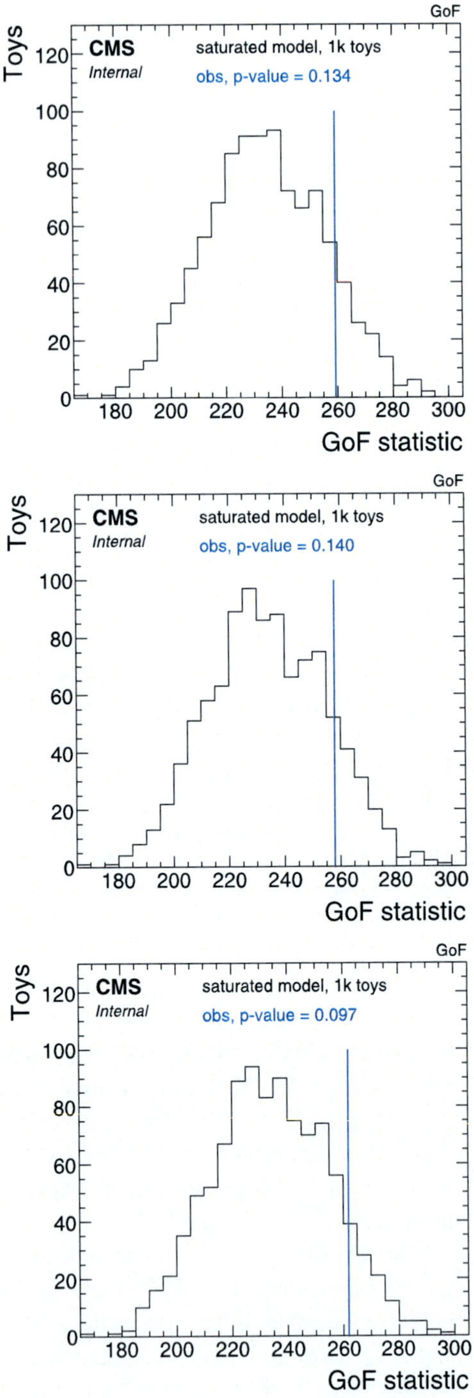

Fig. 9.12 Goodness-of-fit tests between unblinded data and best-fit $\hat{r}S + B$ model. The observed value (obs) of the GoF statistic under the best-fit $\hat{r}S + B$ model (blue line) is compared with the distribution of GoF statistics under the background-only hypothesis (black histogram), for various signal mass hypotheses. Pseudo-experiments over 1k toys are used to generate the distribution in the latter. The p-value for the observation given the background-only hypothesis is quoted for each plot. Upper: $m_{\mathrm{a}} = 100\,\mathrm{MeV}$, Middle: $m_{\mathrm{a}} = 400\,\mathrm{MeV}$, Lower: $m_{\mathrm{a}} = 1\,\mathrm{GeV}$

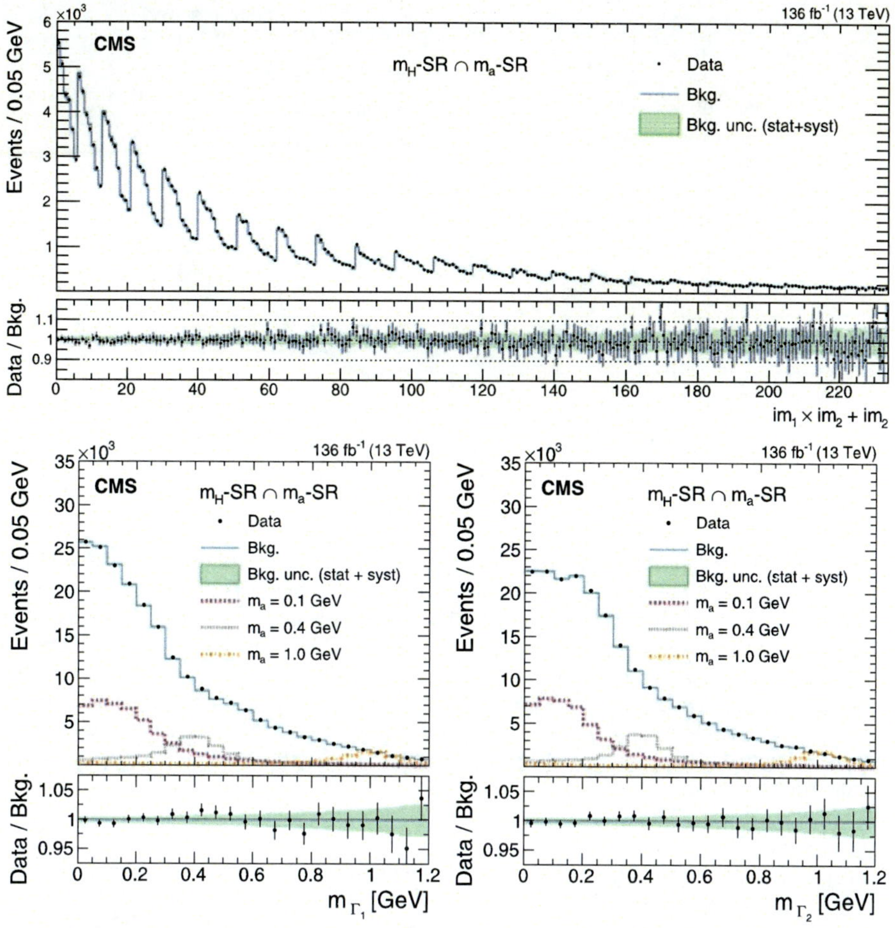

Fig. 9.13 Expected versus unblinded 2D-m_Γ spectra in the final signal region. Upper: unrolled 2D-m_Γ distribution. Lower: projected 1D-m_Γ distributions corresponding to the leading (left) and subleading (right) photon candidate. In the upper panels of each plot, the black points (Date) are the observed data values, with error bars corresponding to statistical uncertainties. The expected background (blue line) uses the optimized values for the systematics strength modifiers determined from the likelihood fit. The total statistical plus systematic uncertainties, also determined from the likelihood fit, are shown as a green band (Bkg unc) around the background model (Bkg). The plotted spectra of simulated $\mathrm{H} \to \mathrm{aa} \to 4\gamma$ events for $m_\mathrm{a} = 0.1$ (purple dashed curve), 0.4 (gray dotted curve), and 1.0 GeV (orange dash-dotted curve) are each normalized to the value of their expected upper limits (Sect. 9.1.5) times 10^3. In the lower panels, the ratio of the observed over background value is shown as black points, with error bars corresponding to statistical uncertainties in the former. The ratio of the total uncertainty over the background model is shown as a green band. Credit: Ref. [7]

The greatest improvement in the limits is most likely to be achieved by employing a dedicated $\mathrm{a} \to \gamma\gamma$ versus SM tagger. While the m_Γ regressor represents an $\mathrm{a} \to \gamma\gamma$ mass reconstruction first, it is not in itself designed for $\mathrm{a} \to \gamma\gamma$ versus background selection. Instead, this role is played by the photon identification criteria of Sect. 6.3,

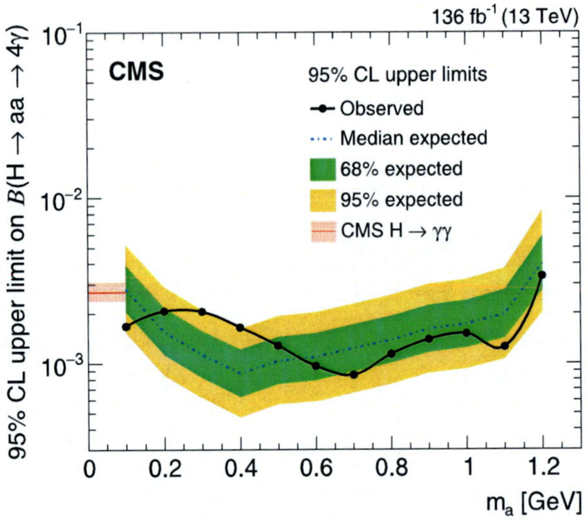

Fig. 9.14 Observed (black solid curve with points) and median expected (blue dashed curve) upper limits on excluded $H \to aa \to 4\gamma$ signal strengths, at the 95% confidence level (CL). The limits are expressed in terms of the total $H \to aa \to 4\gamma$ branching fraction $\mathcal{B}(H \to aa \to 4\gamma)$ and evaluated for the signal mass hypotheses $m_a = [0.1, 1.2]$ GeV, in 0.1 GeV steps. The 68% (± 1 std. deviation) and 95% (± 2 std. deviation) confidence intervals around the median expected upper limits are shown as green and yellow bands, respectively. The 95% CLupper limit corresponding to the CMS $\mathcal{B}(H \to \gamma\gamma)$ measurement [4] is also shown for comparison (red band, where the width represents the uncertainty in the measurement). Credit: Ref. [7]. A tabulated version of this plot can be found in the HEPData record [8]

Fig. 9.15 Results of the signal significance extraction in the unblinded data as a function of m_a hypothesis

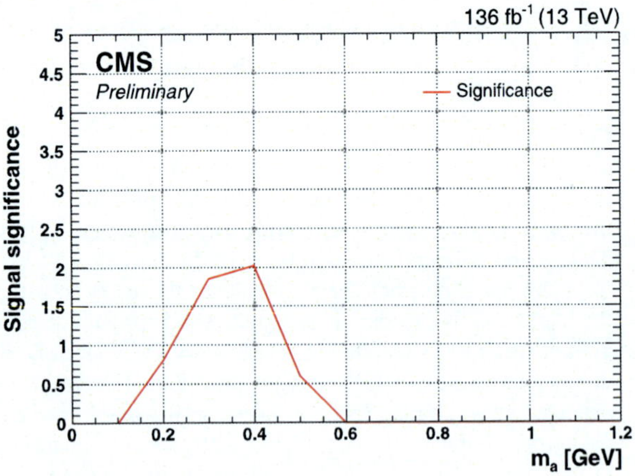

in particular, the EGM photon ID MVA. Since EGM photon ID MVA is optimized for the discrimination of photons versus neutral meson decay from jets, it is sub-optimal for the task of a $\rightarrow \gamma\gamma$ classification, and was simply chosen for simplicity. The development of a dedicated a $\rightarrow \gamma\gamma$ versus SM tagger, potentially one based on the same end-to-end ML principles as for the m_Γ regressor, thus remains an attractive option for significantly improving the sensitivity of this analysis in a future iteration.

9.2.5 Discussion

Given the assumptions made in Sect. 3.3.1 about the H $\rightarrow$ aa $\rightarrow 4\gamma$ phenomenology, particularly those of prompt decay and identical particle a mass, one might ask how robust our results are upon relaxing some of these assumptions. Indeed, long-lived a $\rightarrow \gamma\gamma$ decays are expected to be a general feature of the low-$m_{\rm a}$ regime [6]. While a dedicated search for long-lived a $\rightarrow \gamma\gamma$ would train the m_Γ regressor against a distribution of a lifetimes, even with the m_Γ regressor in its current form, it is worth investigating the regressor's sensitivity to long-lived decays.

We estimate the upper limits for long-lived a decays by comparing the signal yield in the m_H-SR $\cap$ $m_{\rm a}$-SR for different simulated a decay lengths compared with that for prompt decays. The simulation of these long-lived samples is otherwise identical to those of the signal samples described in Sect. 5.2. For $m_{\rm a} = 0.1$ (0.4) GeV, the 95% CLupper limit on $\mathcal{B}$(H $\rightarrow$ aa $\rightarrow 4\gamma$) is 1.6 (0.9) times the prompt-decay upper limit for a distribution of a decay lengths parametrized by $c\tau_0 = 1$ mm, and 30 (3) times larger for $c\tau_0 = 10$ mm, with a linear interpolation between the two limits in both cases. Long-lived decays with displaced vertices will, on average, subtend a smaller angle at the ECAL surface and thus appear more merged. Although this generally makes the a $\rightarrow \gamma\gamma$ more difficult to reconstruct, it does improve photon selection efficiency because the a $\rightarrow \gamma\gamma$ will look more single-photon-like. For most $m_{\rm a}$ and $c\tau_0$ combinations, the net effect is an overall increase (worsening) of the upper limits. However, for $m_{\rm a} = 0.4$ GeV and $c\tau_0 = 1$ mm, the improvement in photon selection efficiency outweighs the poorer mass reconstruction resulting in a slight decrease (improvement) in the upper limits.

To better characterize the impact of the long-lived a $\rightarrow \gamma\gamma$ topology on the regressed m_Γ, we performed studies using simulated, single a $\rightarrow \gamma\gamma$ particle decays, with a range of fixed proper lifetimes (i.e., in the a $\rightarrow \gamma\gamma$ rest frame) $c\tau_0 = 0.07$, 0.14, 0.21 cm. It is important to note that these correspond to samples with fixed $c\tau_0$ rather than the exponential distribution described by parameter $c\tau_0$ of the long-lived H $\rightarrow$ aa $\rightarrow 4\gamma$ signal samples above. We generated particle guns at a fixed mass of $m_{\rm a} = 0.2$ GeV with a fixed energy of $E_{\rm a} = 60$ GeV (roughly the energy if produced from a Higgs boson). This translates to a Lorentz boost of $\gamma_L = E_{\rm a}/m_{\rm a} = 300$, giving a detector-frame lifetime of $cT \approx 20, 40, 60$ cm, respectively. As a reminder, the ECAL is situated at a distance of about $r = 129$ cm from the beamline. As shown in the left plot of Fig. 9.16, the m_Γ regressor loses little sensitivity upto $cT \lesssim 20$ cm, and remains usable upto $cT \lesssim 40$ cm. Beyond that, the angle subtended by the a $\rightarrow \gamma\gamma$

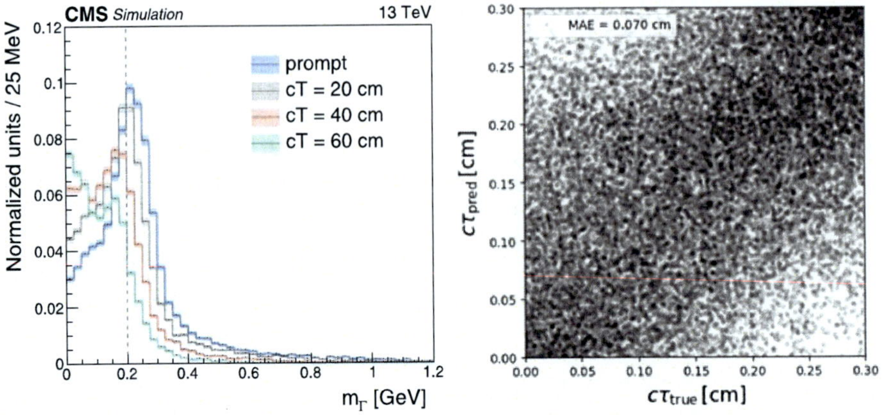

Fig. 9.16 Simulated $a \to \gamma\gamma$ samples with detector-scale lifetimes, at a fixed mass of $m_a = 0.2\,\text{GeV}$ with fixed energy $E_a = 60\,\text{GeV}$. Left: regressed m_Γ for samples that decay promptly (blue) versus those which decay with detector lifetimes of $cT = 20$ (gray), 40 (red), 60 (green) cm. Right: regressed versus generated proper lifetime for long-lived $a \to \gamma\gamma$ samples reconstructed using an end-to-end lifetime regressor. The generated proper lifetimes are uniformly distributed in $c\tau_0 = (0, 0.3)\,\text{cm}$, corresponding to detector lifetimes of $cT = (0, 90)\,\text{cm}$

decay at the ECAL surface becomes too small, and the m_Γ gradually deteriorates into a photon-like m_Γ spectrum. If decaying even closer to the ECAL surface still, the $a \to \gamma\gamma$ begins to enter the ECAL crystals at highly oblique angles, and is more likely to fail the quality criteria for ECAL hit reconstruction.

The above long-lived m_Γ regression results are best understood in terms of the $a \to \gamma\gamma$ lifetime resolution of the ECAL. Using the same end-to-end reconstruction technique used to build the m_Γ regressor, we construct a basic $a \to \gamma\gamma$ lifetime $c\tau_\Gamma$ regressor using simulated $a \to \gamma\gamma$ samples with a continuous distribution of lifetimes, and similar kinematics properties to those described above. Importantly, the $c\tau_\Gamma$ regressor has to be trained with domain continuation on *both* boundaries of the lifetime domain, as physical constraints are present on both the low end (positive-definite lifetime) and high end (ECAL volume) of the lifetime range. We use the same inputs as before (see Sect. 7.1), and find no benefit to including the timing information from the reconstructed ECAL detector hits as an additional image layer[3].

The regressed versus generated proper lifetimes (i.e., in the rest frame of the particle a) are shown in the right plot of Fig. 9.16. This result suggests that, in fact, the proper lifetime resolution of the ECAL for boosted $a \to \gamma\gamma$s is quite poor, with a mean absolute error (MAE) of $\Delta c\tau_0 \approx 0.07\,\text{cm}$. For a Lorentz boost of $\gamma_L = 300$, this translates to a detector-frame decay length uncertainty of about 20 cm. We thus do not expect the m_Γ regressor to be able to discriminate between prompt $a \to \gamma\gamma$ decays

[3] While the difference in path length generated by a displaced $a \to \gamma\gamma$ decay vertex is well below the ECAL single-crystal timing resolution, as was the case for m_Γ regression, it is nonetheless important to test if there are crystal-by-crystal correlations in the timing distribution of the ECAL detector hits that could be exploited by an end-to-end ML algorithm.

from long-lived ones, for as long as the a lifetime is less than the lifetime resolution. Beyond that, we expect a gradual skewing of the m_Γ distribution toward a photon-like spectrum. We note that the deterioration of the m_Γ spectrum is characterized by a gradual loss in efficiency on the left side of the mass peak, rather than an outright displacement of the peak's center. Thus, the primary impact of a long-lived $\mathrm{a} \rightarrow \gamma\gamma$ decay is to reduce the significance of the $\mathrm{a} \rightarrow \gamma\gamma$ signal peak, by increasing the apparent photon-like contribution. Under such a scenario, we expect our background estimation procedure to remain valid. Therefore, *any* $\mathrm{a} \rightarrow \gamma\gamma$ resonance observed in the m_Γ spectrum, whether arising from a promptly decaying a or otherwise, would be indicative of a particle at that mass.

On the other hand, relaxing the assumption of identical particle a masses has a more discontinuous effect on the analysis. For dissimilar a masses, $\mathrm{H} \rightarrow \mathrm{a}_1\mathrm{a}_2$, with $m_{\mathrm{a}_1} \neq m_{\mathrm{a}_2}$, we expects models predicting mass differences less than the m_a-SR window, $|m_{\mathrm{a}_1} - m_{\mathrm{a}_2}| \lesssim 0.3\,\mathrm{GeV}$ to exhibit upper limits comparable to those of prompt decays. However, for larger mass differences, the signal 2D-m_Γ peak will begin to fall outside of the m_a-SR signal region, making a measurement impossible.

To summarize, while we have not optimized directly for long-lived $\mathrm{a} \rightarrow \gamma\gamma$ decays, we expect the interpretation of our results to remain valid even under the long-lived case, although the values of the upper limits are likely to weaken with increasing a lifetime. If dissimilar particle a masses are allowed, our results would only be robust to mass differences approximately less than the $\mathrm{a} \rightarrow \gamma\gamma$ mass resolution at a given mass hypothesis.

References

1. TLHCG (2011) The ATLAS collaboration, the CMS collaboration. In: Procedure for the LHC Higgs boson search combination in Summer 2011. Technical Report CMS-NOTE-2011-005. ATL-PHYS-PUB-2011-11, CERN, Geneva, Aug 2011
2. Cowan G, Cranmer K, Gross E, Vitells O (2011) Asymptotic formulae for likelihood-based tests of new physics. Eur Phys J C 71
3. Gross E, Vitells O (2010) Trial factors for the look elsewhere effect in high energy physics. Eur Phys J C 70:525 Oct
4. Sirunyan AM, Tumasyan A, Adam W, Ambrogi F, Asilar E, Bergauer T, Brandstetter J, Dragicevic M, Ero J, Del Valle AE et al (2019) Combined measurements of Higgs boson couplings in proton-proton collisions at $\sqrt{s} = 13$ TeV. Eur Phys J C 79:421
5. Khachatryan V et al (2014) Observation of the diphoton decay of the Higgs boson and measurement of its properties. Eur Phys J C 74(10):3076
6. Curtin D, Essig R, Gori S, Jaiswal P, Katz A, Liu T, Liu Z, McKeen D, Shelton J, Strassler M et al (2014) Exotic decays of the 125 gev higgs boson. Phys Rev D 90, Oct 2014
7. Search for exotic Higgs boson decays $\mathrm{H} \rightarrow \mathcal{AA} \rightarrow 4\gamma$ with events containing two merged diphotons in proton-proton collisions at $\sqrt{s} = 13$ TeV (2022)
8. HEPData record for this analysis (2022)
9. Collaboration A (2012) Search for a Higgs boson decaying to four photons through light CP-odd scalar coupling using 4.9 fb^{-1} of 7 TeV *pp* collision data taken with ATLAS detector at the LHC 7
10. Adhikari S et al (2022) Search for photoproduction of axionlike particles at GlueX. Phys Rev D 105:052007

Chapter 10
Conclusions

In this thesis, the first CMS search for exotic Higgs boson decays, of the form $\mathrm{H} \rightarrow \mathrm{aa}$, $\mathrm{a} \rightarrow \gamma\gamma$, where each leg is reconstructed as a single photon candidates, is performed. The search is motivated on three fronts: the continued viability of the extended Higgs sector in light of current SM constraints, the exploration of yet unprobed physics phase space due to limitations in current CMS particle reconstruction, and the ability of novel deep learning techniques to make breakthroughs in addressing a number of these reconstruction challenges.

For the first of these, the importance of probing the SM-neutral Higgs sector in light of current allowed BSM physics at the LHC was emphasized. In particular, exotic Higgs decays to a pair of light, axion-like pseudoscalars remains a theoretically well-motivated pursuit, and an important research program for CMS. Below the production thresholds of the heavier fermions, one of the few remaining accessible decay modes is that of $\mathrm{a} \rightarrow \gamma\gamma$, the primary focus of this thesis. The analysis we presented, however, was agnostic to the CP parity of the particle a.

A direct study of the $\mathrm{H} \rightarrow \mathrm{aa} \rightarrow 4\gamma$ spectrum, however, remains elusive due to the extreme collimation of the photons, evading current state-of-the-art particle reconstruction techniques. Yet this remains a critical corner of unexplored physics phase space because of its overlap with existing SM $\mathrm{H} \rightarrow \gamma\gamma$ measurements, within which such a signature may go unnoticed.

By leveraging deep learning techniques, an end-to-end m_Γ regressor was developed that utilizes the ECAL detector data to measure the invariant mass of highly collimated photons reconstructed as a single photon candidate Γ. Together with the development of the domain continuation technique, this thesis has achieved a first in mass reconstruction, allowing particle a masses below the ECAL mass resolution to be reconstructed in simulated $\mathrm{H} \rightarrow \mathrm{aa} \rightarrow 4\gamma$ events.

Using this newfound capability, a search was developed for $\mathrm{H} \rightarrow \mathrm{aa} \rightarrow 4\gamma$ events in the CMS Run II data, corresponding to an integrated luminosity of $136\mathrm{fb}^{-1}$. The two-dimensional merged photon mass, 2D-m_Γ, was used as a discriminant with which to directly identify the $\mathrm{a} \rightarrow \gamma\gamma$ resonance from $\mathrm{H} \rightarrow \mathrm{aa} \rightarrow 4\gamma$ events. In addition, by understanding and controlling the response of the mass regressor to hadronic

M. Andrews, *Search for Exotic Higgs Boson Decays to Merged Diphotons*, Springer Theses, https://doi.org/10.1007/978-3-031-25091-0_10

jets, a total data-driven background model was obtained for all but the SM $H \to \gamma\gamma$ process, which instead relied on simulated events.

The $H \to aa \to 4\gamma$ search was conducted under the framework of a statistical hypothesis test. The MLE method was used to construct a test statistic to determine the compatibility of the data with the $H \to aa \to 4\gamma$ scenario versus the the SM-only one, by extracting the best-fit signal strength of the data to a signal-plus-background model. The CL_s metric was then used to report the signal significance in the event of a detection, or set upper limits on excluded signal strengths, in the event of no detection.

A number of studies were then performed to asses the impact of the different systematics on the best-fit signal strength and the expected upper limits, prior to unblinding. Upon unblinding, the data was found to be consistent with SM-only expectations. BSM models predicting $H \to aa \to 4\gamma$ production with a branching fractions $\mathcal{B} \gtrsim (0.9, 3.3) \times 10^{-3}$ were excluded at the 95% confidence level, for $m_a \in [0.1, 1.2]$ GeV. Based on limited studies of simulated, long-lived $a \to \gamma\gamma$ decays, these results are expected to be robust to particles a with decay lengths $\lesssim 40$ cm, at progressively reduced sensitivity.

Outlook. A major component of this analysis was the development of the m_Γ regressor, and more generally, of the end-to-end particle reconstruction technique [1] with CMS data. In an attempt to manage the risk associated with using such a novel technique on an already novel search, a number of simplifications had to be made in the probed phase space. However, having considerably mitigated these early risks, and having demonstrated the potential of the end-to-end technique, we reflect here on a number of potential improvements that could be implemented in future versions of this analysis, or, indeed, in future spin-offs from this analysis.

With respect to improving sensitivity within the same analysis phase space, as done in most Higgs boson searches, the signal search could be decomposed by Higgs production mode, allowing the higher efficiency modes to bring their full weight into the analysis. This, of course, would involve considerably more work, and would benefit from better integration with the `flashgg` framework. On the other hand, as stated in the Chap. 6, the photon MVA can be replaced with a dedicated $a \to \gamma\gamma$ tagger. While the m_Γ certainly enabled this analysis to be performed in the first place, building dedicated signal taggers have historically allowed drastic reductions in background contamination, with correspondingly dramatic improvements in signal sensitivity. Furthermore, an event classifier, absent from this analysis, could also be developed to further reduce non-resonant backgrounds by exploiting event-level kinematic differences in a more optimized way.

On the other hand, a potential spin-off analysis would be to replace the supervised signal tagger with an unsupervised one, trained directly on data from appropriate sidebands. While this may come at a cost of signal efficiency for the strictly $a \to \gamma\gamma$ signal, it could potentially improve sensitivity to unknown exotic topologies involving merged photons. Indeed, one of the allowed decay modes of the a is to three π^0s, which would be significantly suppressed by our photon identification criteria. This would also open the door to fully relaxing the identical-particle-mass

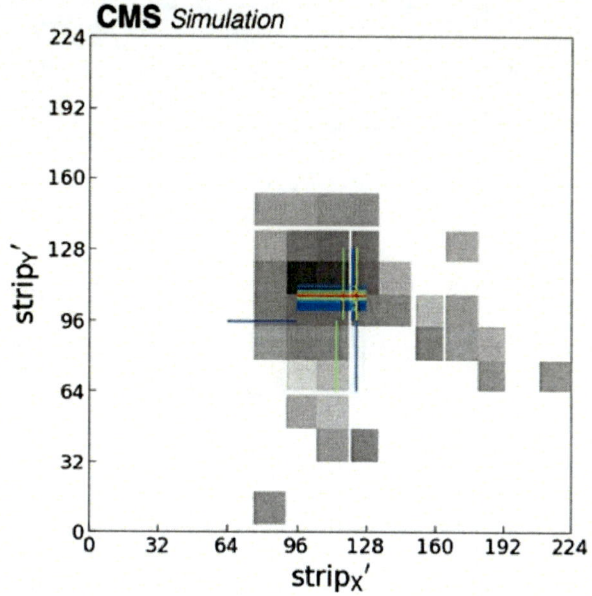

Fig. 10.1 Representative single a $\to \gamma\gamma$ decay in the ECAL endcap. The deposits from the endcap crystals are shown as gray squares, while those from the horizontally-(strip_X') and vertically-segmented (strip_Y') preshower layers are shown as strips in the colored scale

assumption. However, this would likely require more clever ways of scanning over different background regions, each blinded to a different region of the 2D-$m_\Gamma$plane.

Returning to the current analysis, the endcap could finally be included. In Fig. 10.1, we illustrate the potentially dramatic improvement in mass resolution (in the form of angular resolution) that can be achieved by including detector images of the ECAL preshower (ES). With a granularity $\approx 15\times$ that of the ECAL, the addition of the ES would enable highly-boosted a $\to \gamma\gamma$ that are instrumentally merged in the ECAL (gray squares), to be fully resolved in the ES (colored strips).

The m_a phase space could also be extended to higher masses. As noted in Sect. 4, the limiting factor in this case was the deterioration of the $m_{\Gamma\Gamma}$ peak, resulting in significant spill-over into the m_H-SB. This is fundamentally driven by the poorer p_T resolution achieved for loosely merged a $\to \gamma\gamma$ decays in the standard PF-based reconstruction. This is another application that could greatly benefit from the end-to-end particle reconstruction technique. In Fig. 10.2, we show the dramatic improvement achieved by reconstructing the p_T of merged a $\to \gamma\gamma$ decays using an end-to-end p_T regressor (E2E, blue) over a PF-based one (PF, red), and the resulting improvement in the $m_{\Gamma\Gamma}$ mass resolution, in Fig. 10.3. More work, however, would be needed to bring these prototypes into production, and for which reason, they have been left for future work.

Finally, given the lifetime regression discussed previously, a final analysis spin-off could be one that performs a direct measurement of the lifetime of photon candidates. While searches for long-lived particles have been performed at CMS, none have attempted to measure the lifetime spectrum of particles directly. However, given the relatively poor resolution achieved for merged photons, a viable alternate analysis would be a lifetime measurement for H $\to$ aa, a $\to e^+e^-$ or a $\to \mu^+\mu^-$ instead.

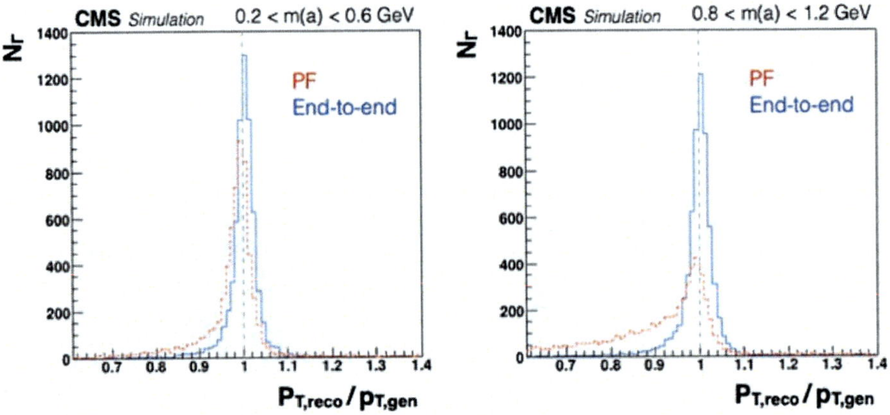

Fig. 10.2 The ratio of the reconstructed over generated p_T of merged $a \to \gamma\gamma$ candidates, using the end-to-end-based p_T regression (E2E, blue), versus the standard PF-based one (PF, red). Simulated single $a \to \gamma\gamma$ samples, with particle masses in the range of $0.2 < m_a < 0.6$ GeV (left) and $0.8 < m_a < 1.2$ GeV (right) are used

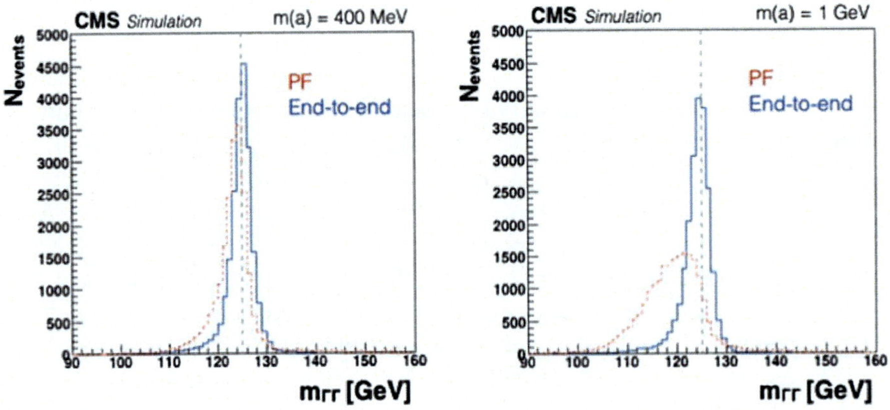

Fig. 10.3 The resulting reconstructed $m_{\Gamma\Gamma}$ resonance corresponding to the Higgs boson, using the end-to-end-based p_T regression (E2E, blue), versus the standard PF-based one (PF, red). Simulated $H \to aa \to 4\gamma$ samples, with particle masses of $m_a = 0.4$ (left) and 1 GeV (right) are used

This would allow track information to be leveraged for potentially better lifetime resolution.

While we have performed a search for $H \to aa \to 4\gamma$, many of the techniques developed to make this analysis possible enable a number of other novel searches to be performed, searches that have not been done because they have been too challenging. Given the spate of null results at the LHC (or at least at CMS [2]), it will become increasingly important to confront such decay topologies that may have been overlooked, due to limitations in existing reconstruction techniques, but for which modern machine learning tools pave a way forward.

References

1. Reconstruction of decays to merged photons using end-to-end deep learning with domain continuation in the CMS detector (2022)
2. Aaij R, Beteta CA, Ackernley T, Adeva B, Adinolfi, Afsharnia H, Aidala CA, Aiola S, Ajaltouni Z, Akar S, Albrecht J, Alessio F, Alexander M, Albero AA (2021) Test of lepton universality in beauty-quark decays

Appendix
Supplementary Studies

A.1 m_Γ spectrum of QCD jets

To better understand the topology and substructure of the QCD jet background and how this affects the response of the m_Γ regressor, the following studies on simulated $\gamma + \text{jet}$ samples are provided. These studies are performed with no trigger requirement and looser event and object selection criteria compared to the main analysis, to maximize the limited simulated data available. In particular, the photon MVA and $\mathcal{I}_{\text{tk}}$ requirements are removed (see Sect. 6.3), and the full $m_{\Gamma\Gamma} > 100\,\text{GeV}$ is used, to enrich the selected sample with jets. For simplicity, MINIAOD rechits are used instead of the nominal AOD rechits used in the main analysis, though the conclusions of this section are not expected to be altered by this choice.

As noted in the QCD substructure discussion in Sect. 8.2.1 but also in the reasoning for using $\text{Z} \to \text{e}^+\text{e}^-$ electrons to validate the regressor in Sect. 7.5.2, at the energies relevant for the Higgs regime ($E \sim 60\,\text{GeV}$), the approximation that the $\pi^0 \to \gamma\gamma$s passing the photon identification criteria are well-isolated begins to break down. As shown in Fig. A.1, the photon isolation properties of EM-enriched jets passing photon ID, are distinctly busier than those of similar mass $\text{a} \to \gamma\gamma$ decays which are more inline with those of true photons decays (e.g. $\text{H} \to \gamma\gamma$ or even γ production). This arises because $\pi^0 \to \gamma\gamma$ are fundamentally produced as jets from QCD and at higher energies additional constituents in the jet are produced. The shower shape and isolation cuts contained in the photon ID will then preferentially select jets which are collimated. Because of this collimation, we do not *a priori* expect to be able to regress clean $\pi^0 \to \gamma\gamma$ peaks. In fact, as the energy of the jet increases, more energy becomes available to create additional $\pi^0 \to \gamma\gamma$s in the jet. Furthermore, because of the photon ID shower shape requirements, we will preferentially be selecting such jets where these multiple $\pi^0 \to \gamma\gamma$ decays are all very closely merged together in a single cluster. Thus, not only do we expected the $\pi^0 \to \gamma\gamma$ peak to be distorted, we expect it to drift upward with increasing p_T as the effective mass of the jet grows. That is, we expect the m_Γ spectrum to be positively correlated with p_T.

M. Andrews, *Search for Exotic Higgs Boson Decays to Merged Diphotons*, Springer Theses, https://doi.org/10.1007/978-3-031-25091-0

To illustrate these effects, we first focus on the subleading photon candidate in the $\gamma+$jet sample. Because the neutral meson faking the photon typically only has a fraction of the parent parton's energy, it is usually the subleading photon candidate that contains the meson. In Fig. A.2, we plot the regressed $m_{\rm a}$ (left column) and reconstructed $p_{\rm T}$ (right column, labelled $p_{\rm T,a}$) for the subleading photon candidate (black curves). From this distribution, we can then require the photon candidate to be ΔR-matched ($\Delta R < 0.04$) to a generator-level neutral meson decay (blue curve): either a $\pi 0$ (upper row) or an η meson (lower row). Compared to the corresponding spectrum for a $\rightarrow \gamma\gamma$ decays (see for instance Fig. 4.1), the $m_{\rm a}$ spectrum for the π^0-matched sample (upper row, blue curve) exhibits a thicker peak but also a broad hump at around $m_{\rm a} \approx 500$ MeV that is not due to the η meson. We observe the resonance from the η-matched sample (lower row, blue curve), to be subject to similar distortion effects. With a heavier mass of $m_\eta = 550$ MeV and a much lower production cross section, however, the η will appear to be even less prominent because of these distortion effects. We do note, however, that, upon closer inspection (lower row, middle plot), a well-isolated η component is visibly regressed as a sharp mass peak.

To illuminate the π^0-matched spectrum further, we can additionally require that the photon candidate be $p_{\rm T}$-matched ($p_{\rm T}^{\rm reco}/p_{\rm T}^{\rm gen} < 1.1$) to the generated π^0 to which it was already ΔR-matched. This gives us an approximate handle[1] for discriminating between jets containing single $\pi^0 \rightarrow \gamma\gamma$s (likely to be $p_{\rm T}$-matched) versus jets with multiple $\pi^0 \rightarrow \gamma\gamma$s (likely to have over-estimated reconstructed $p_{\rm T}$). In Fig. A.3, we plot only the ΔR-matched component (black curve) of the subleading photon candidate $m_{\rm a}$ (left column), reconstructed $p_{\rm T}$ (middle column), and $p_{\rm T}^{\rm reco}/p_{\rm T}^{\rm gen}$ (right column). These curves are then decomposed into their p_T-matched (blue curves) and non-p_T-matched contributions (red curves). The decomposed π^0-matched sample (upper row) clearly shows what looks like a sharp a $\rightarrow \gamma\gamma$-like peak (blue curve) and a broad $m_{\rm a} \approx 500$ MeV spectrum (red curve) unique to the jets. As the red curve corresponds to the over-estimated reconstructed $p_{\rm T}$ tail (right column), we identify this broad spectrum as coming from jets of multiply-merged $\pi^0 \rightarrow \gamma\gamma$s. Decomposing the η sample (lower row) shows that this effect is much less pronounced for the heavier η, consistent with our earlier observations. While the broad $m_{\rm a} \approx 500$ MeV spectrum of the π^0-matched sample sits close to the the true η-meson peak, the shape is distinctly different, suggesting the regressor realizes this component of the QCD spectrum is out-of-sample—as it is. Indeed, even the photon ID MVA realizes this, as it is this component of the jet that is classified as most "fake-like" (score ≈ -1) and is the first to be removed by cutting on the photon ID MVA. However, care must still be taken when attempting to claim an η observation in data.

If indeed this broad $m_{\rm a} \approx 500$ MeV spectrum is due to multiple merged $\pi^0 \rightarrow \gamma\gamma$s as we have claimed, we would further expect its spectrum to be $p_{\rm T}$-dependent: as the jet's energy grows, so, too, does its potential to produce more Π^0s, thereby increasing

[1] The truth table of generated particles in the MINIAOD dataset (as is used in this study) truncates soft particles below $p_{T,\rm gen} < 10$ GeV so it is not possible to directly, reliably count how many generated π^0s are in the vicinity of the reconstructed photon.

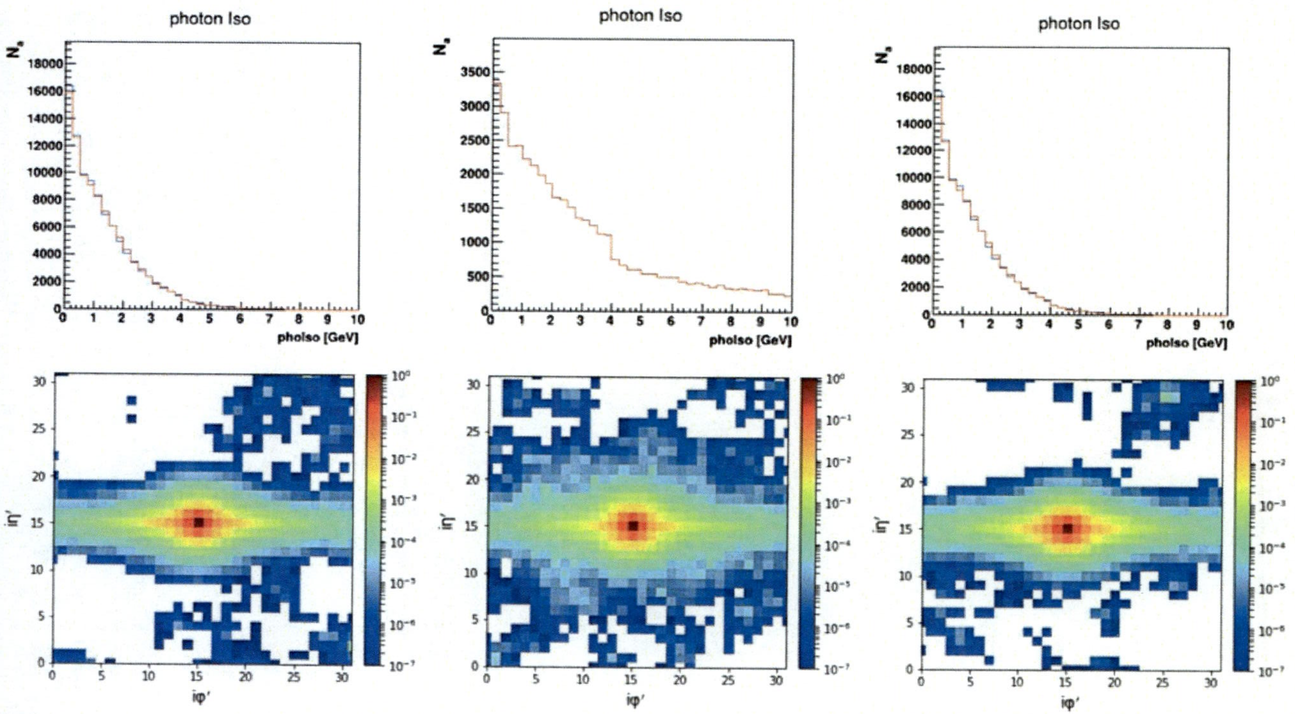

Fig. A.1 Photon isolation distributions (upper row) for subleading photon candidate representing EM-enriched QCD jets (middle column) and a $\rightarrow \gamma\gamma$ decays at nearby masses $m_a = 100$ MeV (left column) and $m_a = 200$ MeV. Corresponding ECAL overlays shown in lower row, as normalized to most energetic crystal. QCD jets typically contain additional neutral (among other) constituents translating to more pronounced tails in their photon isolation distribution

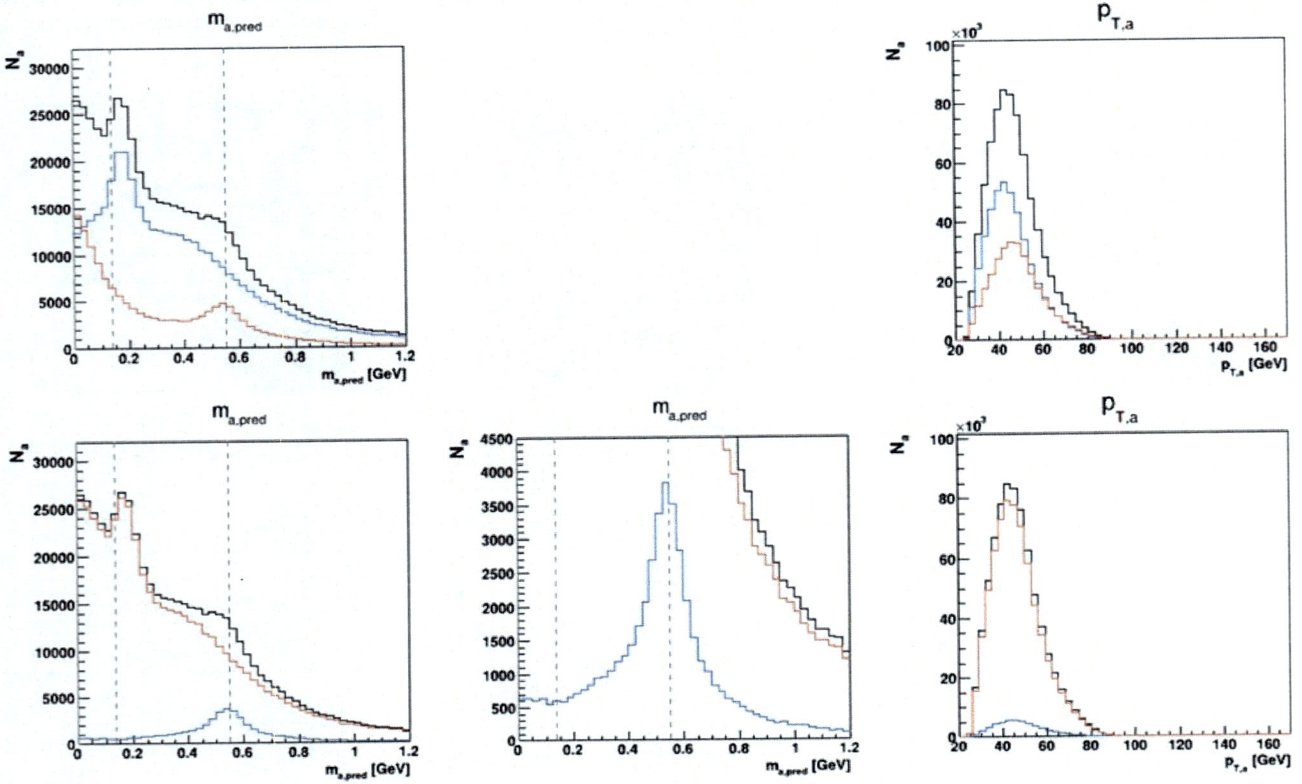

Fig. A.2 Regressed m_{a} (left and middle column) and reconstructed p_{T} (right column) of subleading photon candidate in γ + jet events. Black curves show total distribution, blue curves show component ΔR-matched to a π^0 (upper row) or to an η meson (lower row) and red curves show non-ΔR-matched component. Lower middle plot shows η-matched inset of lower left plot to emphasize m_{a} peak

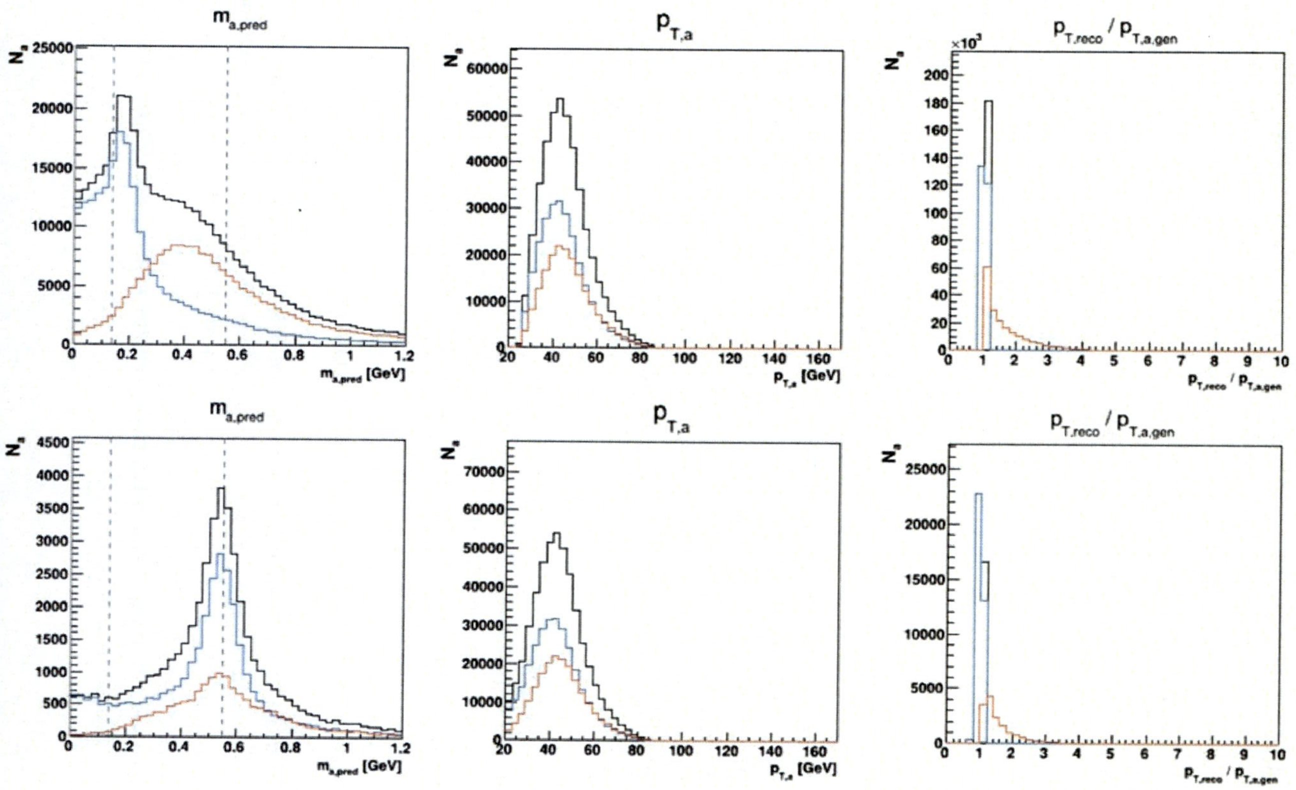

Fig. A.3 Regressed m_{a} (left column), reconstructed p_{T} (middle column), and $p_{\mathrm{T}}^{\mathrm{reco}}/p_{\mathrm{T}}^{\mathrm{gen}}$ (right column) of subleading photon candidate in $\gamma+$jet events. Black curves show distributions ΔR-matched to a π^0 (upper row) or to an η meson (lower row). Blue curves show p_{T}-matched component appearing most $\mathrm{a}\to\gamma\gamma$-like and red curves show non-p_{T}-matched component containing multiple merged neutral mesons, a distinct feature of QCD jets the mass of the jet. We can verify this explicitly by plotting the regressed m_{a} versus the reconstructed p_{T}. These are shown in Fig. A.4 for the total ΔR-matched sample (left column) and its decomposition into p_{T}-matched (middle column) and non-p_{T}-matched (right column). The π^0-matched sample (upper row) confirms this: the p_{T}-matched component (middle plot) has the same stable regressed mass as seen for a $\rightarrow \gamma\gamma$s [1] (within the approximate decomposition that $p_{\text{T}}^{\text{reco}}/p_{\text{T}}^{\text{gen}}$ allows) whereas the non-p_{T}-matched sample has the positive correlation we expect for multiple $\pi^0 \rightarrow \gamma\gamma$s. The η-matched sample shows similar features, although to the lesser degree that we expect.

As a final check, we can also look at the leading photon candidate. While this is expected to contain less jets, for the few that it does, we do expect to see an enhancement of this broad $m_{\text{a}} \approx 500$ MeV component, as, with higher p_{T}, we have more energy to produce multiple $\Pi^0 \rightarrow \gamma\gamma$s and are less likely to create a single isolated $\Pi^0 \rightarrow \gamma\gamma$. In the upper row of Fig. A.5, we plot the regressed m_{a} (left plot) and the reconstructed p_{T} (right plot). The total leading m_{a} spectrum (black curve) is, as noted, much more photon-enriched with the component ΔR-matched to Π^0s (blue curve) much less dominant. For completeness, the non-ΔR-matched component (red curve) is also plotted and is essentially photon-like. To get a better view of the jet spectrum, in the lower row, we focus in on just the Π^0-matched component (now plotted as the black curve) and, as before, decompose this into its p_{T}-matched (blue curve) and non-p_{T}-matched component (red curve). Indeed, the broad $m_{\text{a}} \approx 500$ MeV spectrum seen previously is much more dominant in the leading candidate, just as predicted. This can also be seen in the more pronounced over-estimated reconstructed p_{T} tail (lower row, right plot). At this energy, the regressor has difficulty regressing even the clean, single $\Pi^0 \rightarrow \gamma\gamma$-like component of the jet (blue curve) although this is consistent with a $\rightarrow \gamma\gamma$s of high-p_{T} and low-m_{a} and is to be expected given the level of photon merging.

The above analysis confirms our assessment that QCD jets at this energy have unique features not present for the signal a $\rightarrow \gamma\gamma$ topology but which are nonetheless reasonably interpreted by the m_Γ regressor, given the underlying physics. Of course, these still need to be accounted for when building the background model. Specifically, the p_{T}-dependent nature of the QCD jet m_Γ spectrum must be corrected for, as the m_H-SB events used to build the background model exhibit a different p_{T} spectrum from those observed in the m_H-SR. This, therefore, culminates in the two mitigation strategies implemented in this analysis: first, cutting down the p_{T}-dependent component of the jet via a photon ID MVA and $\mathcal{I}_{\text{tk}}$ cut (and for which reason they were left out of this analysis) as given in Sect. 6, and second, applying a p_{T} re-weighting procedure to correctly capture the p_{T} spectrum of the m_H-SR events, as described in Sect. 8.2.2.

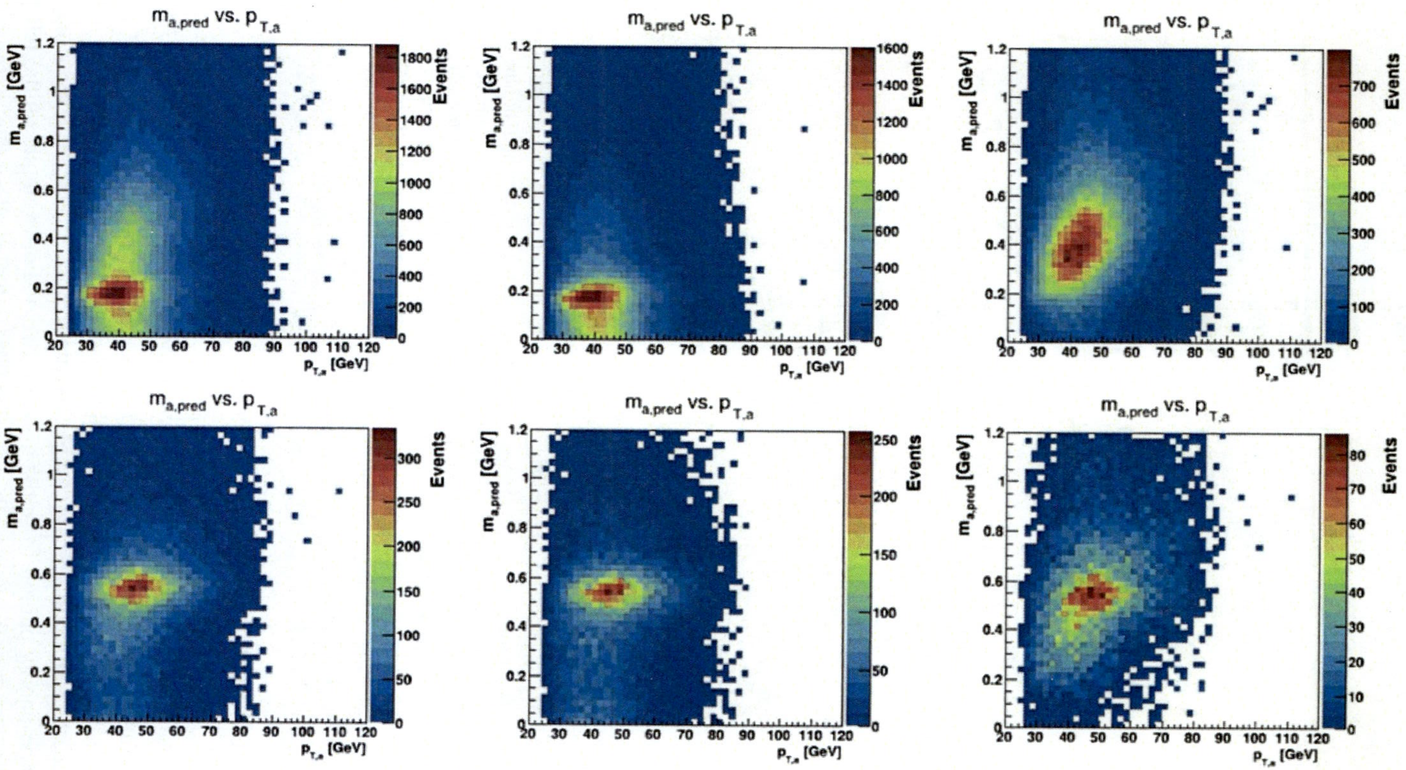

Fig. A.4 Regressed m_{a} vs. reconstructed p_{T} distribution of subleading photon candidate in γ + jet events. Left column represents candidates ΔR-matched to a π^0 (upper row) or to an η meson (lower row). Middle column shows only the p_{T}-matched component of these, showing flat, a $\rightarrow \gamma\gamma$-like response while right column shows only the p_{T}-matched component showing p_{T}-dependence unique to QCD jets

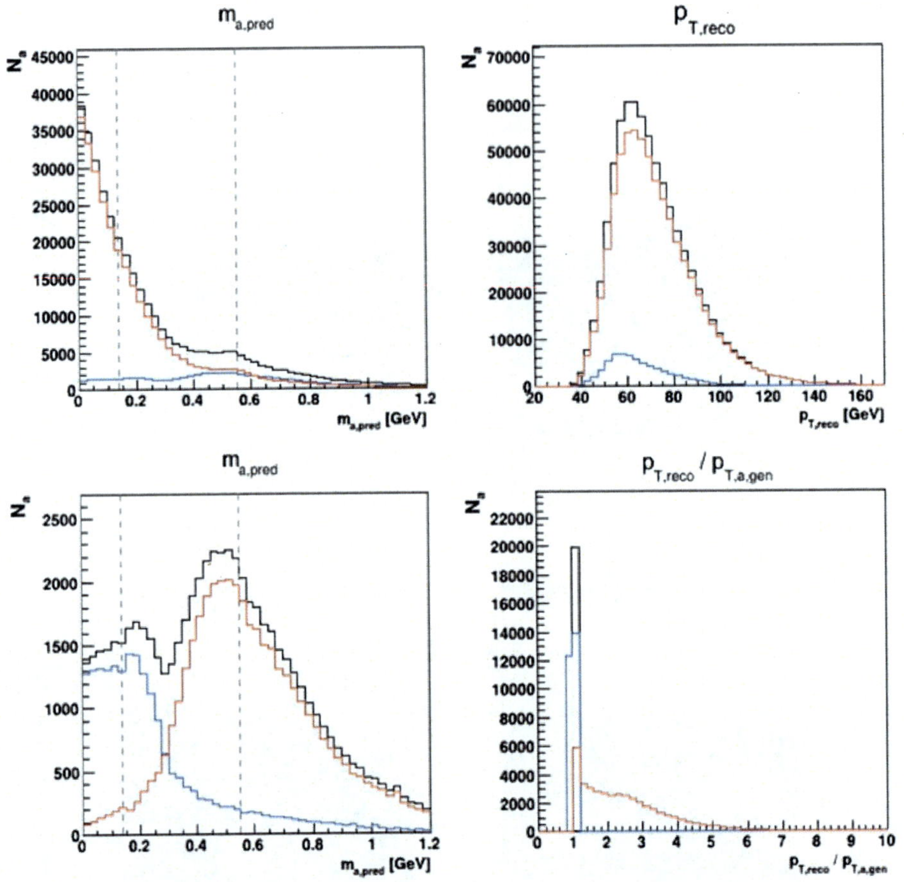

Fig. A.5 Upper row shows regressed m_a (left column) and reconstructed p_T (right column) of leading photon candidate in γ + jet events. Black curves show total distribution, blue curves show component ΔR-matched to a Π^0, and red curves show non-ΔR-matched component. Lower row shows only ΔR-matched distribution for m_a (left column) and p_T^{reco}/p_T^{gen} (right column) in inset. Black curves in lower row correspond to ΔR-matched curves plotted in blue in upper row. Also in lower row, blue curves show p_T-matched component appearing most a $\rightarrow \gamma\gamma$-like and red curves show non-p_T-matched component containing multiple merged neutral mesons which are enhanced at higher p_T

A.2 Minimally Processed Versus Clustered Data

We attribute the robustness of the end-to-end ML mass regressor (see Sect. 7.5) to the use of minimally processed (*all*) rather than clustered (*clustered*) detector data. As shown in Fig. A.6, the PF clustering algorithm filters out low-energy deposits and, under certain situations, may completely remove all the deposits associated with the lower-energy photon from the a $\rightarrow \gamma\gamma$ decay. For example, on the left plot of Fig. A.6, showing the minimally processed data, the lower-energy photon is visible on the lower left of the core photon, at a distance of approximately 7 ECAL crystals.

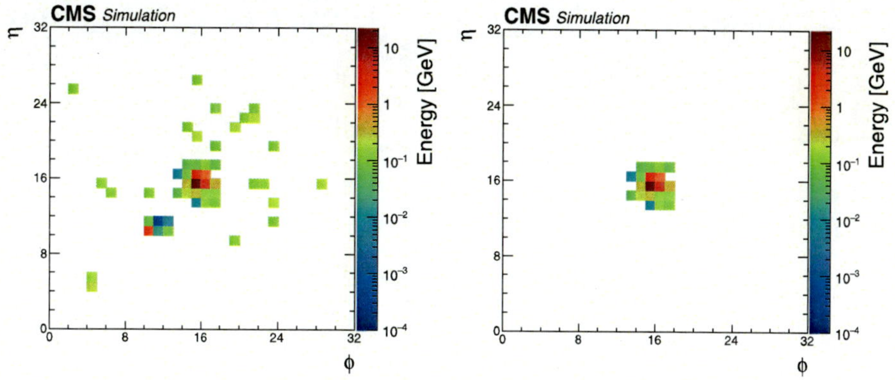

Fig. A.6 The azimuthal angle ϕ versus the pseudorapidity η of a typical $a \to \gamma\gamma$ decay using minimally processed (left) and clustered (right) data

In the clustered data on the right plot, the deposits associated with the lower-energy photon have been dropped, along with other, isolated low-energy deposits.

The impact of using minimally processed versus clustered data can be seen in Fig. A.7. We compare the effect of training on all (upper row) versus clustered data (lower row), for shower-merged boosts (left column) and barely-resolved boosts (right column). For each scenario, we reconstruct the mass of a sample constructed from all (blue circles) versus clustered (red squares) data, to compare how well each mass regressor extrapolates to the other's domain. For the mass regressor trained on minimally processed data (upper row), despite the differences in input image (c.f. Fig. A.6), we see no evidence of a shift in the mass peak in either boost regime. This is not the case for the mass regressor trained on clustered data (lower row), which exhibits a shift when applied outside of its domain. This suggests it is desirable not to filter the detector data beforehand so that the mass regressor learns to suppress low-energy deposits. Without this opportunity, the mass regressor becomes more susceptible to variations in the data.

In addition, the loss of the lower-energy photon at resolved boosts (right column) leads to $a \to \gamma\gamma$ decays being incorrectly reconstructed as photons, causing both a drop in reconstruction efficiency at the correct mass peak ($m_\Gamma \approx 1$ GeV) and a build-up of samples at the wrong peak ($m_\Gamma \approx 0$ GeV). Even if minimally processed data is presented to the mass regressor trained on clustered data (lower right, blue circles), the lost efficiency at the correct mass peak is still not recovered. The use of minimally processed data is thus vital to maximizing the capabilities of the end-to-end ML-based technique.

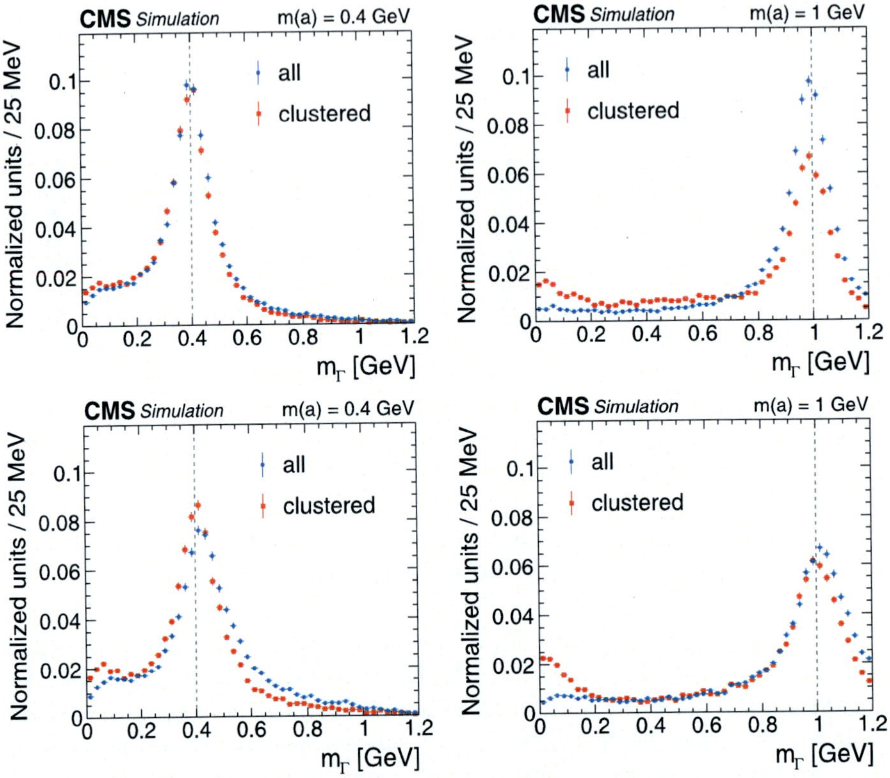

Fig. A.7 Predicted mass spectra for the mass regressor trained on minimally processed data (upper) versus clustered data (lower) at shower-merged boosts (left) and barely resolved boosts (right). For each scenario, the mass regressor is run on the same set of a $\rightarrow \gamma\gamma$ decays, composed either of minimally processed (blue circles, all) or clustered (red squares, clustered) data. The vertical dotted line shows the input m_a value

Reference

1. Reconstruction of decays to merged photons using end-to-end deep learning with domain continuation in the CMS detector (2022)